ELEMENTS OF GRAPHICS

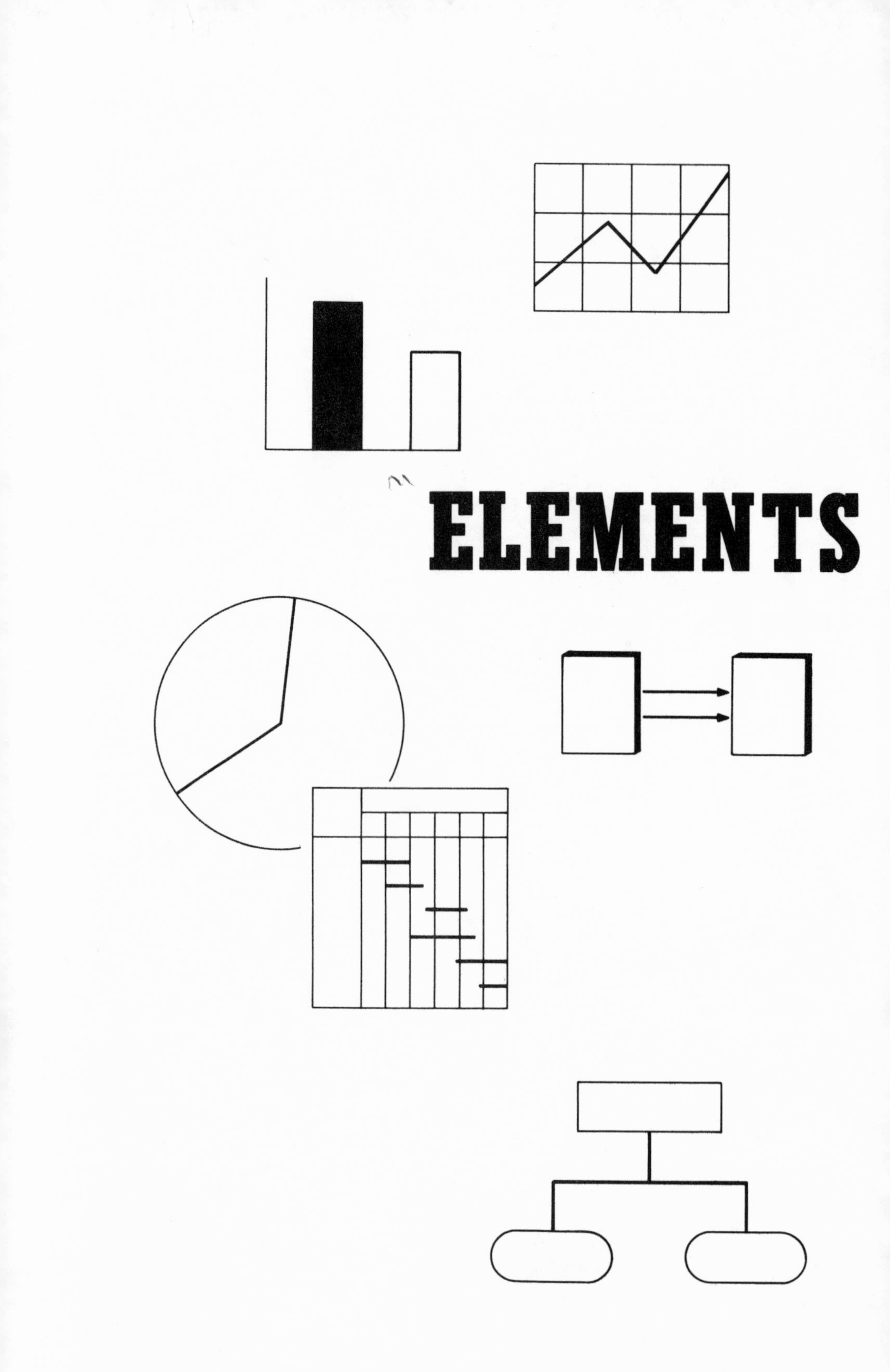
ELEMENTS

ROBERT LEFFERTS

OF GRAPHICS

how to prepare charts and graphs for effective reports

1817

HARPER & ROW, PUBLISHERS, New York,
Cambridge, Hagerstown, Philadelphia, San Francisco,
London, Mexico City, Sào Paulo, Sydney

 For information address Harper & Row, Publishers, Inc., 10 East 53rd Street, New York, N.Y. 10022. Published simultaneously in Canada by Fitzhenry & Whiteside Limited, Toronto.

Designer: Ruth Bornschlegel

Library of Congress Cataloging in Publication Data

Lefferts, Robert.
Elements of graphics.
Includes index.
1. Graphic methods. I. Title.
QA90.L413 001.4'226 80-8209
ISBN 0-06-012578-0

82 83 84 85 10 9 8 7 6 5 4 3 2

Contents

Preface

Communication through written reports has become a major characteristic of modern society. The achievement of social, political, and economic goals increasingly depends upon our ability to present information in a clear, concise, and persuasive manner. Charts and other graphic devices are effective ways to complement narrative reports in order to enhance their power of explanation and persuasion. This book provides an introduction to how to prepare charts and other graphics so that these methods can be accessible to all report writers.

After many years of working with students and a variety of professional persons, I am convinced that any report writer can master the preparation of graphics and other illustrations. Most people like illustrations and visual images. Thus, they already have a sense of graphics. What is needed is an explanation and a demystification of graphic methods. In a society whose economic system is organized to inhibit the sharing of expertise it is a constant struggle to increase the availability of knowledge for the use of all persons. This book is a small contribution to that struggle. It contains all of the information necessary to prepare high-quality, effective graphic illustrations.

I am indebted to Cathy Trainor and other graduate students in my systems analysis courses who first suggested that I put in book form the material they were studying on how to devise and use graphic illustrations. I appreciate the cooperation of the administration of the Health Sciences Center of the State University of New York at Stony Brook, Dr. J. Howard Oaks and Dr. Daniel Fox, for facilitating my work on the book. Rick Lefferts, a planner, and Herbert Wainer shared helpful information with me. Rick Balkin of the Balkin Agency provided sound guidance and encouragement. Reginald Bromeen was helpful in ways that defy description.

A principal contributor to this effort has been Sybil Lefferts, whose professional ideas, judgments, and criticisms are reflected throughout the book.

The staff of the Health Sciences Center, Media Services Medical Illustration Department, Kathy Gebhart, Karen Henrickson, and Tom Giacalone faithfully prepared camera-ready illustrations with good humor and professional expertise.

Finally, I would like to acknowledge the superb assistance of Susanne Torjussen, whose secretarial skills, encouragement, and suggestions are seen on every page.

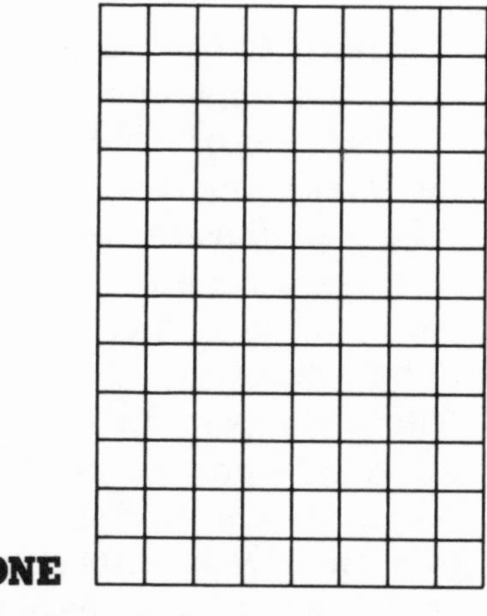

CHAPTER ONE

Introduction

This book shows how to prepare and use charts and other graphic methods to make written reports more effective. The basic principles and skills involved in the preparation of graphics are outlined with step-by-step directions that enable the nonartist to devise professional-looking and technically correct graphic illustrations. The book is concerned exclusively with the kinds of charts, graphs, and pictorial representations that can be used in everyday nonscientific reports prepared by managers, administrators, government officials, professionals, university students, and others. These graphic illustrations include bar charts, pie charts, line charts, organization charts, flow charts, time-line charts, and various graphic techniques to create additional interest and emphasis.

There are literally millions of reports written every year on an almost endless number of subjects. All are intended to convey information and to persuade some particular audience of the importance of what is being presented. Many report writers, however, fail to achieve the full impact of their purpose because of the absence of graphics to supplement and complement the words and numbers that comprise their reports. In what has been called the "communication age" every report writer is competing for the attention of audiences who are being bombarded with written and aural messages. Graphics not only attract attention but also present material in a clearer and more interesting manner.

The preparation and use of graphics is viewed by many writers as depending on methods and techniques that are the exclusive prerogatives of professional artists and illustrators, engineers, draftsmen, or statisticians. As a result, most report writers have been mystified into believing that preparing graphics is beyond their abilities. They are intimidated by the belief that they "can't draw a straight line."

The purpose of this book is to provide the principles, methods, and examples that will make graphic presentation accessible to any report writer. With the use of easily available material and the simple instructions in the book, high-quality graphics can be prepared by writers who do not have the benefit of specialized art departments or professional illustrators. All these graphics can be prepared by hand or on the typewriter, in black and white.

The book is not intended as a substitute for the more sophisticated artwork necessary in highly technical reports or for the kind of illustration and layout involved in publishing, packaging, advertising, or the like. Rather, all of the graphic methods in the book are of the kind that are suitable for inclusion in typewritten reports and that can be reproduced using the various types of duplicating machines available to most organizations and offices.

The specific objectives of the book are to enable report writers to develop an understanding of:

- What is a graphic?
- Why do graphics make a report more effective?
- What are the attributes of a good graphic?
- When and where should you use a graphic illustration?
- What are the various types of graphics?
- How can you prepare a good graphic?
- What are the materials you need to assist you in preparing graphics?

The book is directed to a variety of report writers and, therefore, includes illustrations and examples from a number of different fields. All these can be adapted for your own purposes. The principal groups of report writers who will find the material helpful include:

- persons who occupy a wide range of management and middle-management positions in governmental, nonprofit, business, commercial, service, or industrial organizations
- faculty and students in universities and colleges
- officials in schools, health, education, and human-service agencies and community organizations
- planners, researchers, consultants, and persons in the professions

The preparation of well-designed graphics is both an art and a skill. There are many different ways to go about the task, and readers are urged to develop their own approaches. Graphics can be creative and fun. At the same time, they require a degree of orderly and systematic work. This book is intended to help persons refine and develop their own skills in this field, since no uniform or universally accepted set of guidelines exists that applies to all forms of graphic presentation.

The presentation of the book begins, in Chapter 2, with a discussion of what graphics are and why they are so effective. This includes an explanation of the general principles of good graphics: unity, balance, contrast, and meaning. The effective use of graphics involves awareness of the specific report-oriented objectives that can be achieved through graphics. Chapter 3 explains and illustrates how graphics can be used to accomplish the objectives of clarity, simplification, emphasis, reinforcement, summarization, interest, coherence, or impact. Chapter 4 provides a brief description of the basic materials you will need to prepare your charts.

Chapters 5 to 10 are devoted to instructions on how to use and prepare specific kinds of graphics. These include the three main types of graphics for presenting quantitative or numerical information: bar and column charts (in Chapter 5), pie charts (in Chapter 6), and the line chart (in Chapter 7). A number of methods for presenting qualitative information are then presented, including organization charts (in Chapter 8), flow charts (in Chapter 9), and time charts (in Chapter 10).

In Chapter 11 a number of techniques for adding interest and emphasis to a report are illustrated. Finally, Chapter 12 describes a variety of resources and reference materials that can be utilized to enhance the preparation of graphics. A fuller explanation of the supplies and how they can be used is also included.

Each of the chapters on graphic methods includes a number of examples and illustrations as well as a discussion of:

- What can be achieved through the use of that particular method.
- What kinds of information can be best communicated by the method.
- A step-by-step set of instructions on how to prepare the particular graphic method.
- Common mistakes to avoid in the use of the method.

Graphic illustrations are no substitute for good writing. They can, however, enliven the written material, add clarity, create impact, and enhance the efficiency and effectiveness of a report. What one seeks in a graphic is much like what one seeks in life—stimulation of the senses, unity, balance, meaning. These principles are examined in the next chapter.

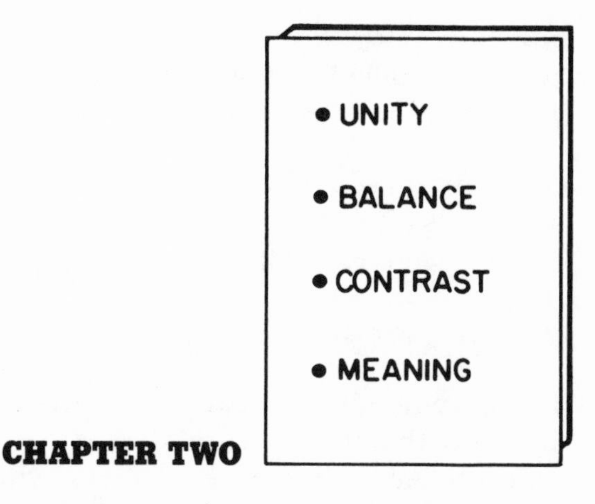

CHAPTER TWO

Principles of Graphics

Graphic charts are ways of presenting quantitative as well as qualitative information in an efficient and effective visual form. Numbers and ideas presented graphically are often more easily understood, remembered, and integrated than when they are presented in narrative or tabular form. Descriptions, trends, relationships, and comparisons can be made more apparent. Less time is required to present and comprehend information when graphic methods are employed. As the old truism states, "One picture is worth a thousand words."

In order to test these assertions, read each of the narrative examples in Chapter 3 without looking at the accompanying chart. Then look at the chart and ask yourself if the chart increased the *efficiency* (by requiring less time), the *effectiveness* (by enhancing your understanding of the information), and the *impact* (by strengthening your impression) of each presentation.

What is it about graphics that gives them these qualities? Examining this question will help us understand when and how to employ these methods most productively. Two subquestions must be answered to gain some insight into the broader issue of why graphics work so well. These are:

- What are graphics and how do they contribute to effective presentation?
- What are the attributes of good graphics?

WHAT ARE GRAPHIC CHARTS?

The graphic charts with which this book is concerned are drawings that use lines and shapes to represent numbers and concepts. As such they represent a synthesis of ancient and modern methods—the methods of prehistoric peoples, of Plato and Aristotle, of the computer; a synthesis of philosophy, geometry, statistics, and art. In graphics, lines and shapes are combined into forms called charts: bar charts, pie charts, organization charts, flow charts, time charts, and others.

These forms can serve as powerful aids to our ability to perceive the world around us and to communicate with others. Graphic forms help us to perform and influence two critical functions of the mind: the gathering of information and the processing of that information. Graphs and charts are ways to increase the effectiveness and the efficiency of transmitting information in a way that enhances the reader's ability to process that information. Graphics are tools to help give meaning to information because they go beyond the provision of information and show relationships, trends, and comparisons. They help to distinguish which numbers and which ideas are more important than others in a presentation.

In the chapters that follow we shall point out (a) how bar charts, pie charts, and line charts can be used to convey quantitative information; and (b) how flow charts, organization charts, time charts, and others convey qualitative information. One bar, for example, may be larger than another. The comparisons are clear to the reader. One box in an organization chart is at the top. The graphic image of hierarchy is simple and clear. The mind can grasp these points with ease and also will retain the information more readily.

Lines and shapes such as circles, rectangles, triangles, and curves have been used by people throughout history to help explain the nature of their environment. These geometric forms have been employed to represent important philosophical views: the circle to represent unity; the arrow to represent attraction, flow, and direction; the rectangle to represent area or mass. That forms can become political and social symbols is evidence of the power of graphics. The star, the swastika, the hammer, the cross—all have been used for political and social purposes.

Rudolf Arnheim has shown that all thinking is basically perceptual in nature, and that the dichotomies between thinking and seeing, between reason and perception, are false and misleading.[1] To see is to reason. Thus, the use of visual forms of communication has great potential for influencing what a person thinks. Graphic presentation is always much more than a way to present just facts or information. Rather, it is a way to influence thought, and, as such, graphics can be a powerful mode of persuasion.

What then does one "see" when looking at a graphic chart? One sees an objective form with a particular shape (for example, round) and one also sees certain subjective ideas (for example, unity, bigger). The more complex the shape of any object, the more difficult it is to perceive it. The nature of thought based on the visual apprehension of objective forms suggests, therefore, the necessity to keep all graphics as simple as possible. Otherwise, their meaning will be lost or ambiguous, and the ability to convey the intended information and to persuade will be inhibited.

An important property of a graphic is the extent to which the observer, in our case a reader of a report, is able to distinguish it from other visual stimuli. Most important, one wants the reader to distinguish it from words. The shapes used in graphic presentation are different from the shapes of letters. And, while graphic charts must be labeled with words, these should be kept to a minimum in order to preserve the effec-

[1]Rudolf Arnheim, *Visual Thinking.* Berkeley: University of California Press, 1969.

tiveness of the graphic presentation. In addition, the more the graphic is distinguished from the "field" (that is, the paper and words) within which it appears, the more the reader's perception is increased. Thus, the arrangement of shapes, the use of heavy dark lines, and adding shading all make the graphic stand out from its field. The reader is helped to distinguish it and, therefore, to pay more attention to it.

WHAT ARE THE ATTRIBUTES OF GOOD GRAPHICS?

Good graphics have certain specific attributes that make them potent instruments of presentation. Understanding them will help you plan and utilize effective graphic presentations in your reports. You will find that these attributes represent things that you already know. To have them further elaborated and to have your own awareness validated should contribute toward the further development of your confidence and skills in the use of graphics.

A graphic is an illustration that, like a painting or drawing, depicts certain images on a flat surface. The graphic depends on the use of lines and shapes or symbols to represent numbers and ideas and show comparisons, trends, and relationships. The success of the graphic depends on the extent to which this representation is transmitted in a clear and interesting manner.

The lines and shapes that comprise a graphic all occupy a common space. Good graphics combine and control these lines and shapes in such a way that they have four attributes:

- unity
- balance
- contrast
- meaning

While these are also principles of art in general, when applied to graphic illustrations they take on a more limited and special meaning that is described in the following sections.

■ Unity

A good graphic must give the impression that its various parts all belong together. They must be arranged in such a way that the illustration looks like a single entity. A good graphic chart should be more than just the sum of its individual lines, shapes, and shades. It should be more than the individual bars in a bar chart, more than the pieces of a pie chart, more than the boxes in a flow chart. Unity requires the establishment of coherent relationships among the component parts of the drawing. These relationships can be depicted in a very direct manner through the use of connecting lines that serve to connect shapes.

This is readily seen in Examples A and B in Figure 2-1

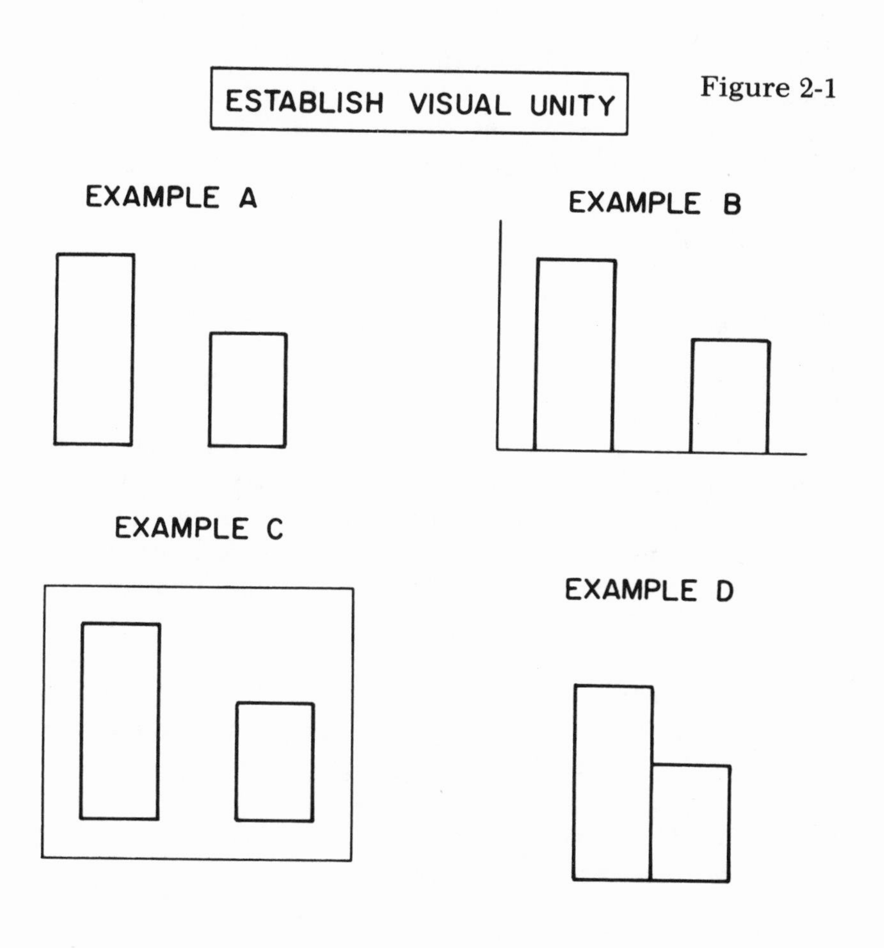

Figure 2-1

showing two rectangles. Example A is not much more than two boxes. Examples B and D, however, represent something more than just two boxes because the boxes are connected by lines, as in B, or are adjacent, as in D. Just what they are intended to represent, however, is not yet clear because certain additional principles of graphic presentation have not yet been incorporated into the drawing: namely, *scales of measurement* and *descriptive labels.* The visual unity, however, has been established.

Do relationships always need to be expressed through such direct connection? No, not always, as in Example C in Figure 2-1. There are times when unconnected shapes, simply because they are similar (both rectangles) and appear on the same plane together with a shared boundary (the box that encloses them), become unified. The organization of lines, where they start and stop, their thickness, their direction, and the arrangement of shapes constitute the basic substance for the visual images we call graphics. Your graphics will be more aesthetically pleasing and substantively persuasive if they use lines and shapes in a way to convey this sense of unity and coherence.

■ Balance

Unlike some art forms, good graphics should be as concrete, geometrical, and representational as possible. A rectangle should be drawn as a rectangle, leaving nothing to the reader's imagination about what you are trying to portray. The various lines and shapes used in a graphic chart should be arranged so that it appears to be balanced. This balance is a result of the placement of shapes and lines in an orderly fashion.

To be orderly, the parts of the chart must be arranged according to a systematic, consistent set of principles. Graphics are very controlled, rigid, and geometric pictorial representations. Random or haphazard organization of the parts of a graphic are not desirable.

Balance in relation to graphic charts is accomplished when two requirements are met:

1. The center of the graphic can be located or visualized. The center is the point where the horizontal and vertical lines do or would bisect each other.

2. The various elements of the graphic (lines and shapes) are more or less equally balanced to the right and left (and/or above and below) of the visual center.

Balance, in this context, refers to a sense of equal weight among the components of the graph; between the two sides; between top and the bottom. Balance is, of course, easily accomplished when two equal things are being described or compared. This is called *symmetrical balance.* But in graphics we are seldom working with equally weighted factors, and we must achieve *asymmetric balance.* Such balance is achieved by planning and controlling the location, size, and arrangement of the lines, shapes, symbols, shadings, scales, and labels that comprise the illustration.

Graphics that are used to represent quantitative information such as bar charts, pie charts, and line charts all rely on the use of vertical and horizontal axes for their basic format, as illustrated in Example A of Figure 2-2. These axes designate the measurements used in the chart. The vertical (or y axis or ordinate) is usually employed to represent an amount or frequency. The horizontal (or x axis or abscissa) is usually used to designate the method of classification that is used. More often than not, the negative ($-x$, $-y$) elements are not necessary and only positive (+) values are being expressed. Therefore, most graphic charts will only actually show the positive format, as in Example B in Figure 2-2

Balance and unity are, of course, interdependent. In constructing graphic charts, one can seldom be achieved without the other. Keeping your charts as simple as possible will add to their clarity and help you to achieve the necessary balance and unity.

■ Contrast

Two aspects of contrast are important to recognize. One is the need to have the graphic contrast with its field. In the

Figure 2-2

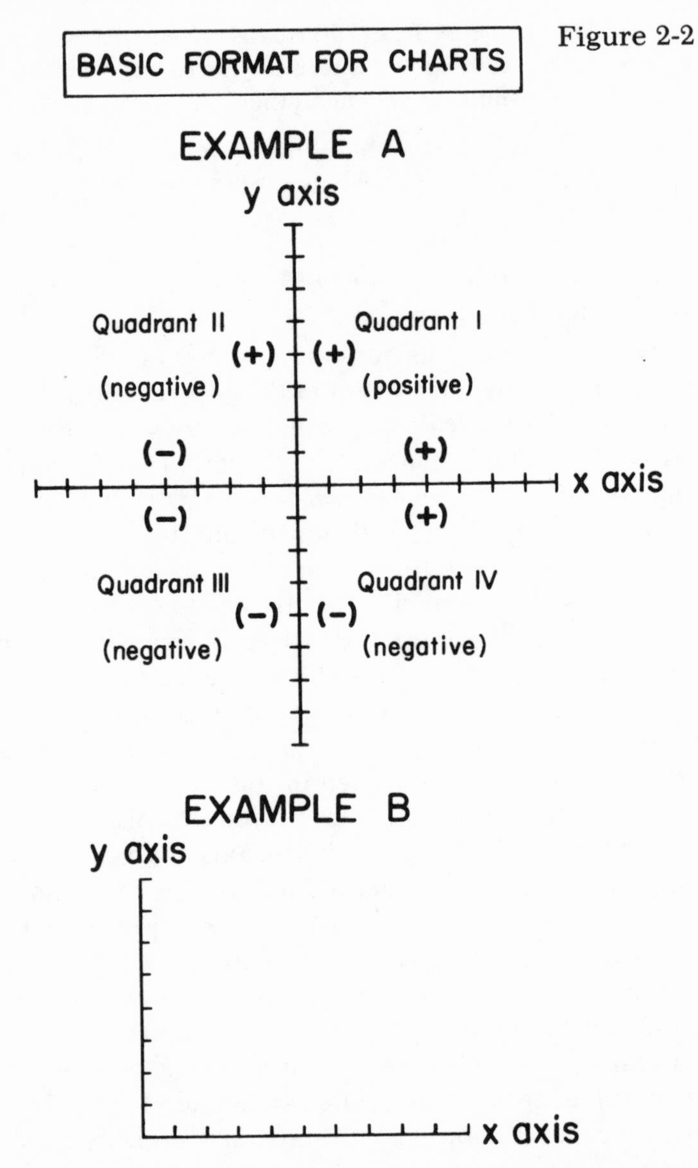

case of reports, this field is the white page of the report. The second aspect of contrast is the need to achieve it within the chart itself.

A graphic is interesting to the extent that (a) it is different from other visual stimuli in a report, such as words and the page itself; and (b) it emphasizes some comparison and provides some insight into the differences or relationships between or among two or more factors. These factors may be quantities, as in bar, pie, and line charts; or qualities, as in flow charts and organization charts. Contrast is accomplished by varying the size, shape, shading, and location of lines and symbols. Some illustrations of this can be seen in Figure 2-3.

Figure 2-3

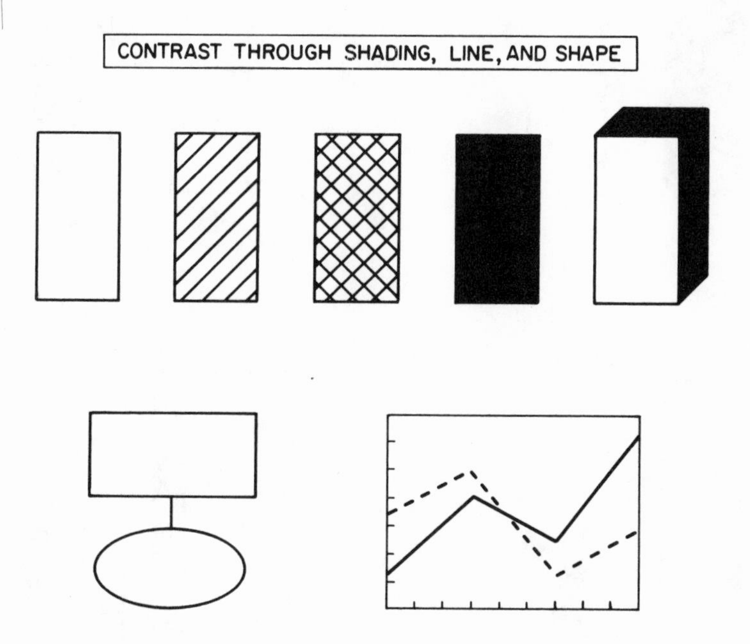

Contrast also depends on establishing relative relationships. Contrast may range from considerable to minor, but there can be no contrast unless some visual difference is established.

Contrast may be real or illusionary. For example, a rectangle can be drawn or shaded so that it appears to have an additional dimension of volume or depth, even though it is on a flat page. In Figure 2-3 the five bars all represent the same quantities. Yet, through the use of shading, they convey

Figure 2-4

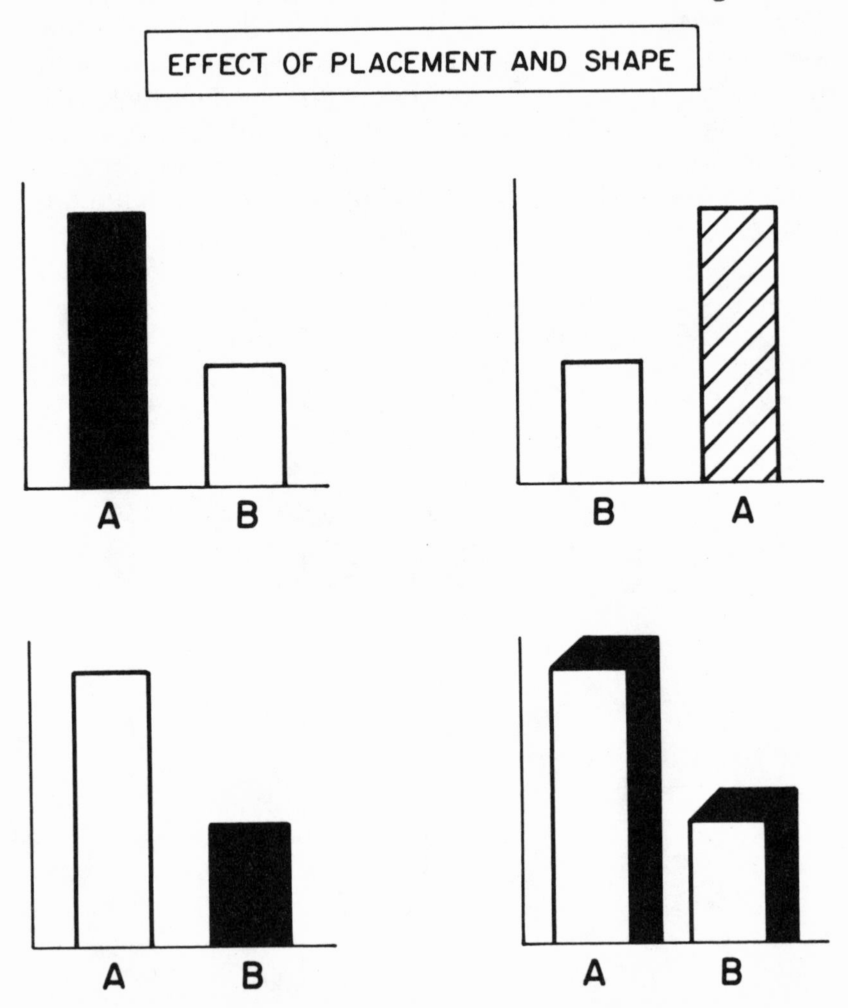

a different sense of importance and size. One of the major issues in the preparation of graphic charts is the extent to which the graphic pictorial contrast distorts the "real" figures or information upon which the chart is based. These distortions can be the result of the way scales of measurement are constructed or they can result from use of lines, shapes, and shading.

In Figure 2-4 there are four bar charts, all of which represent the same data. They show two factors: *A* representing

50% and *B* representing 25%. Quantitatively, A is twice as large as B. However, the extent of the difference between A and B can be manipulated and emphasized in a number of ways. Making one bar darker, changing relative positions, using hatching, and creating additional dimension are all ways to change the visual appearance. Shading, for example, creates a sense of increased volume and size. In addition, darkening or shading makes the object seem closer and makes it stand out more, thus communicating an increased sense not only of proportion but also of importance.

In addition, one should also be aware of the effect of different kinds of lines. For example, a vertical line implies more

Figure 2-5

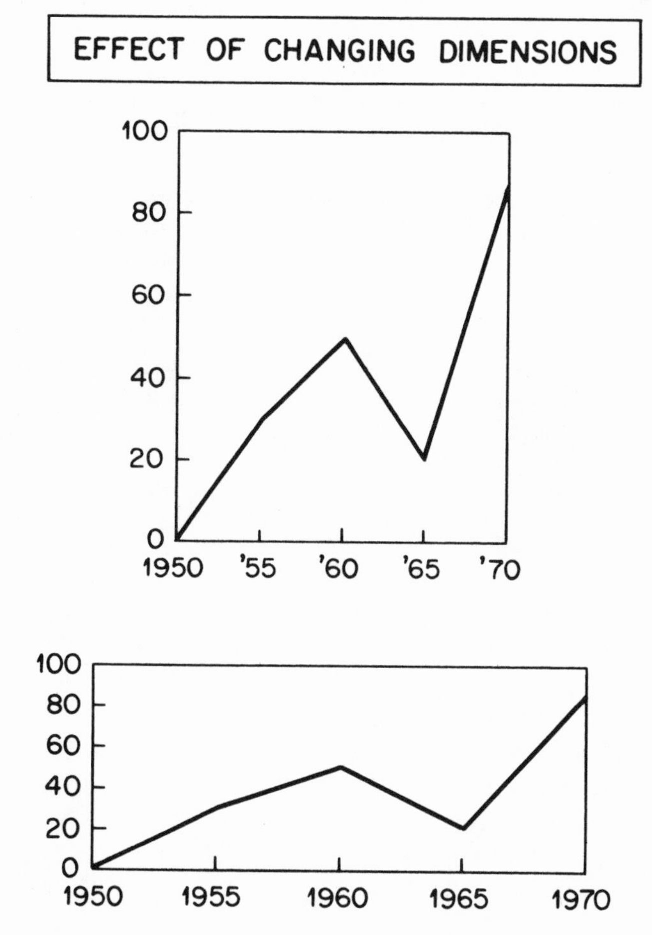

strength than a horizontal line. A diagonal line implies more movement or action than vertical or horizontal lines. In the chapters that follow we shall illustrate various ways to create contrast and additional interest without distortion for each type of chart. We shall also point out some kinds of distortion that should be avoided. One of these is the manipulation of measurement scales. Examine, for example, Figure 2-5, and you will see how changing the distance (or the unit of measure) on the axes of a line chart can appear to change the meaning of two sets of identical data.

In using the thickness of lines, shapes, and shading to achieve contrast, you must exercise caution not to overuse these techniques, or the chart will lose its sense of unity and balance.

■ Meaning

A graphic chart combines lines into certain forms and shapes intended to communicate information in a way that also gives that information certain meanings. In terms of graphic communication, meaning is a combination of what the reader comes to know from the graphic *and* how the reader *perceives* what he or she knows. To attach meaning to a graphic always requires the reader (a) to focus attention on one thing as being more important than another; and (b) to come to understand relationships, trends, and patterns.

Meaning is the result of a translation process in which one sees a certain form or image (the graphic) comprised of lines, shapes, and words; labels it as a circle, box, column; and attaches a definition, understanding, or explanation of what is seen, such as "big," "little," "nice," "important," "good," "confusing," "A is bigger than B," "A is going up every year."

Understanding is accomplished through: (a) the use of relative size of the shapes used in the graphic; (b) the positioning of the graphic-line forms; (c) shading; (d) the use of scales of measurement; and (e) the use of words to label the forms in the graphic. In addition, in order for a person to attach meaning to a graphic it must also be simple, clear, and appropriate. In the following chapters we shall discuss how to prepare graphics consistent with all of these requirements.

There are good graphics and there are poor graphics. A good graphic has the attributes we have been reviewing: unity, balance, contrast, and meaning. The first step in the use of good graphic charts in reports is to develop a clear strategy for your graphic presentation. You can do this by identifying the specific objectives you want to achieve in the report and deciding how graphic illustrations can help achieve those objectives. In the next chapter, nine objectives that can be achieved by graphics are discussed and illustrated. In subsequent chapters, a variety of specific types of graphics are covered.

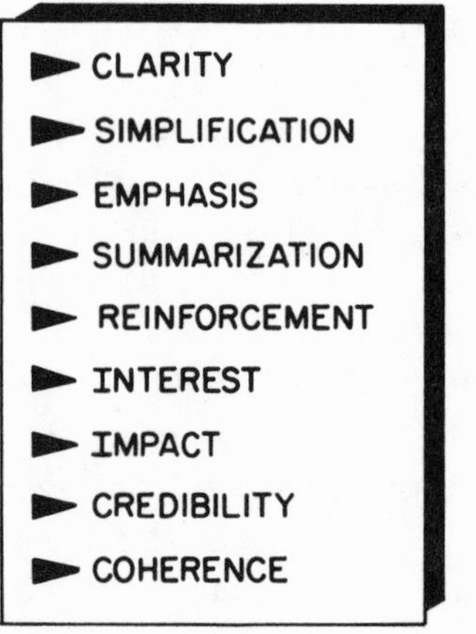

CHAPTER THREE

The Nine Uses of Graphics

Writers must have a clear strategy in mind when they decide to use graphics in a report. It is imperative to be aware of both the general purposes and the specific objectives to be accomplished through graphic illustrations.

Written reports of the kind we are concerned with all share two general purposes or goals: *explanation* and *persuasion.* These purposes mean that the report must convey information that is both understandable and convincing. The reader must not only know what is being said but must also believe in the correctness of the points that are made. The problem facing all report writers is that of finding ways to accomplish these goals. In part, of course, this has to do with the content of the material selected, the writing style, the use of documentation, the overall organization of the report, and the development of a logical argument. Graphics can be an important strategical addition. While they cannot substitute for good writing, they can be a powerful tool to help attain the report's purposes. This is true because graphics can be used to achieve a number of specific objectives, all of which serve to enhance the descriptive and analytical material in a report.

Understanding these objectives is necessary in order to develop a graphic strategy for the report. Such understanding also makes it possible to decide when and where to use graphic material and what kind of graphic to use.

There are nine interrelated objectives that can be achieved through the use of graphics. They all, in their own way, contribute to attaining the goals of explanation and persuasion. These nine objectives are:

- ▶ CLARITY
- ▶ SIMPLIFICATION
- ▶ EMPHASIS
- ▶ SUMMARIZATION
- ▶ REINFORCEMENT
- ▶ INTEREST
- ▶ IMPACT
- ▶ CREDIBILITY
- ▶ COHERENCE

Each of these objectives is examined in the remaining sections of this chapter, and examples are provided of how graphic presentations can be used to achieve the objectives. It should be noted that a number of the objectives are interrelated and overlapping. For example, clarity and simplification are interdependent in the sense that the accomplishment of one results in the accomplishment of the other. In graphics when we make things more clear we also simplify. When we make something more simple we also make it more clear.

Similarly, the objectives of emphasis and reinforcement are interrelated in that any technique to emphasize an idea or information also results in reinforcement of that idea. Conversely, whenever some point is reinforced with the use of a graphic, that point is also emphasized. In like manner, when interest is added to a report it creates greater impact; and to achieve impact through the use of graphics makes the report more interesting. Nevertheless, a writer must be aware of the primary objective desired because some techniques are more suitable than others in terms of each specific objective. Thus, it is helpful to examine each objective separately in order to understand the variety of things that can be accomplished through the use of graphics.

▶ CLARITY

Without clarity it is impossible to create understanding. The reader of a report must be able to identify the specific points that are being made in a presentation and must also be able to grasp the significance of those points. Often, narrative presentation alone is too vague, too cumbersome, or too complicated to present certain material with sufficient clarity to assure accurate communication from the writer to the reader.

Narrative text by itself may also obscure a point or may imply an unwanted ambiguity. Graphics can make the material more clear, and thus more understandable. This is especially true in the case of:

- quantitative or numerical information
- the explanation of changes or trends over time
- presentation of a series of different points
- explanation of the sequence or chronology of events
- showing the relationships among a number of different points, factors, events, activities, processes, or organizational units.

It should be noted that graphics for the purpose of clarity should not be a substitute for words and numbers in the narrative text. The graphics presentation is used to supplement the narrative; otherwise, there wouldn't be anything to clarify.[1]

You must be careful not to try to show too much in a graphic presentation. Only the more complex material that

[1]There are cases when a graphic, particularly a chart, can be usefully employed to present information not included in the narrative text, but the primary objective in doing this is for purposes of brevity or coherence.

you are afraid the reader may not understand in narrative form should be converted to a graphic form in order to achieve clarification.[2] In addition, of course, you need to make a judgment about the material's importance. Do you want to stress certain complex material? If the answer to this question is "yes," then the use of a graphic is appropriate. The graphic enables one to present the material in an alternative form—the pictorial illustration—that readers often find less confusing than words and numbers alone. Following is an excerpt from a report and an example of how a graphic supplement, in this case a *pie chart,* can be used to clarify what some readers might consider an involved quantitative description:

> The Gross National Product of the United States in 1975 was $1,516 billion. For the fifth time the GNP exceeded $1 trillion. The GNP in 1975 was comprised of economic activity in a number of different sectors. Of the total economic product, $54 billion was in the agricultural sector and another $37.6 billion in mining. Together, these two areas, in which our city is most interested, accounted for $92.4 billion or 6.1% of the total. Wholesale and retail trade was $272.4 billion or 18% of the total. The areas of finance, business, and real-estate activity accounted for $209.4 billion, which was 13.8% of the total. The personal-service industry was responsible for $181.8 billion or 12%. Government's direct expenditures were $200.6 billion, which was 13.2%. Manufacturing, the single largest area, accounted for $346 billion, and this was 22.8% of the GNP. Transportation and communication sectors were $95 billion or 6.3%. The remaining 7.8%, some $118 billion, was divided among a number of other miscellaneous areas of economic activity.

The use of a graphic illustration, such as the pie chart in Figure 3-1, serves to clarify this description by selecting the main point—that is, the percent of each sector—for illustra-

[2]Graphics can be used for simple material, of course. However, the objective in these cases is other than clarity; it may be to create emphasis or add impact.

tion. The probability that the reader will comprehend the material more fully is also increased through the use of this illustration—it serves the purpose of simplification as well as that of clarity. In the next section the use of graphics for simplification is discussed further.

Figure 3-1

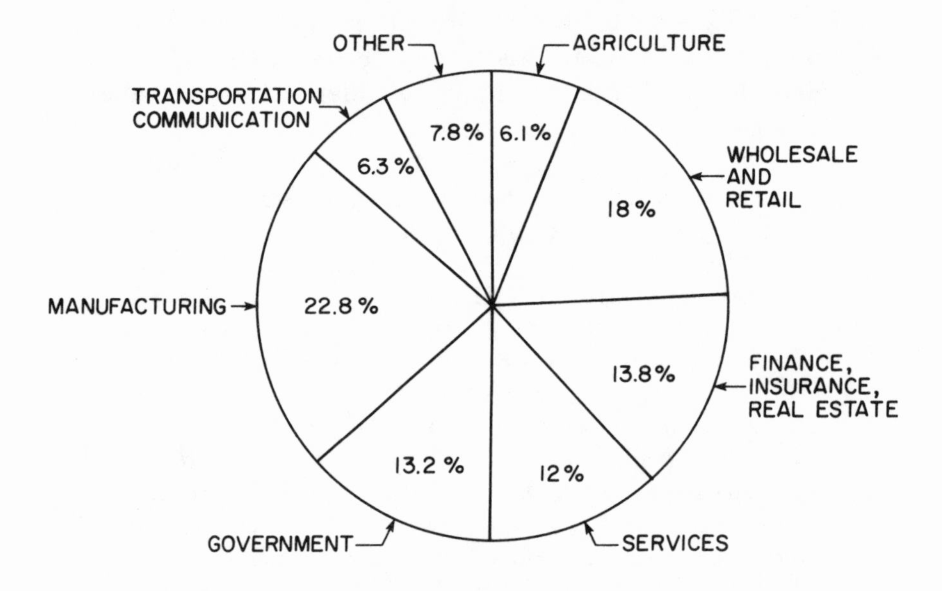

▶ SIMPLIFICATION

As more and more is known about the world around us, and as we become more specialized and technologically sophisticated, our attempts to communicate with each other become more and more difficult. We have more to say, more history to refer to, more qualifying arguments to include, more to describe. Further, readers increasingly want to know the "why" of a particular point. Thus, it becomes necessary to use more words and numbers to make a number of different points, to point out connections, and to use more technical or specialized language. This is as true for explaining nuclear reactors as it is for discussing the contents of a breakfast cereal, as true for discussing schizophrenia as for reporting the Monday-night football game.

A result of the increasing amount and complexity of information that must be presented is that readers are often unable to comprehend sufficiently what the writer is trying to say. Many writers shun the notion of simplification on the basis of "it's just not that simple," or "don't be simplistic," or "it's dangerous to oversimplify."

To simplify through the use of graphics, however, is not to oversimplify or to be simplistic or to take away from the true richness or complexity of the subject. Rather, the function of graphics is to present complexity in a pictorial form in order to complement a narrative explanation. The graphic form itself, because of its structure and pictorial quality, serves to simplify the narrative complexity and make it more understandable to the reader.

Effective simplification is based primarily on the process of partialization—that is, the ability to break a complex whole down into its component parts and yet preserve the whole.

Figure 3-2 is an example of how the following rather complicated set of ideas is broken down into its component parts and presented in a *flow chart* that accounts for each component.

> Management by objectives (MBO) is used extensively in business and industry and increasingly in governmental agencies. The MBO system is simple in theory, but quite complicated in practice. This is because there are many steps and substeps in the implementation of MBO in large organizations.
>
> A prerequisite for successful implementation of MBO is the commitment of top management to the system. Assuming this commitment is there, a process of convincing personnel through the organization of its benefits is required. Management must be prepared to allocate sufficient resources to engage in the various activities needed to implement such a system. These activities require a good deal of internal analysis aimed at defining overall organizational objectives and subobjec-

tives, and doing the same thing for each organizational unit, cost, or profit center.

MBO requires a considerable amount of design activity that uses the information generated during the early analysis stage as a basis for design of forms, reporting schedules, data acquisition, storage and retrieval methods, and plans for how the data will be utilized in decision-making.

Implementation of an MBO system in the organization requires identifying a unit responsible for the management of the MBO system, training personnel in its use, and providing feedback on its performance. An MBO should not be viewed as a one-time process; it should be periodically redefined, based on such feedback, by management, and the process should be repeated to refine the MBO system further.

The inclusion of the chart to complement the foregoing narrative not only presents the material in a simplified form, but also adds to the narrative itself by highlighting the main points for the reader.

THE MBO INSTALLATION PROCESS Figure 3-2

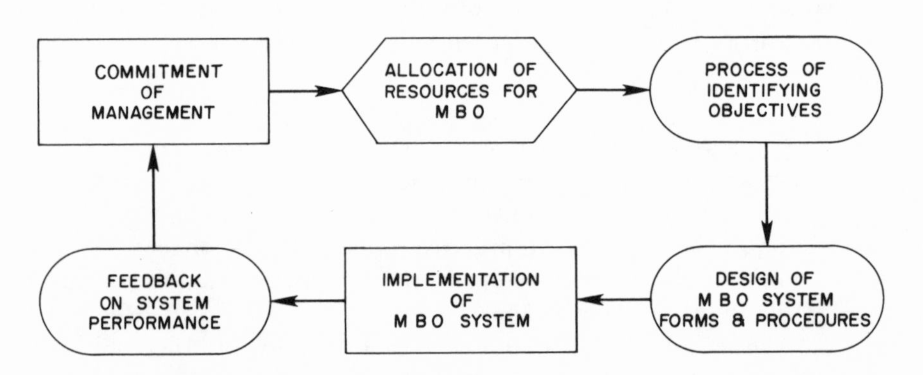

▶ EMPHASIS

Descriptive and analytical reports, of necessity, often cover many different points. But not all the points covered are of equal importance. Report writers usually want the reader to pay more attention to some points than to others. They want certain things to stand out more than other things. Writers recognize that readers will not remember everything that is presented in a report. But there are certain important things they want the reader to remember. Thus, they want these things to be emphasized. In the narrative text this objective is generally accomplished through the use of adjectives, adverbs, and other word choices.

Graphic techniques provide report writers with a set of additional tools to create the emphasis they desire. This is true because the material selected for graphic presentation is visually more prominent than the other material. The reader's attention is effectively called to this selected material.

Suppose, in the example regarding the GNP presented earlier in this chapter, the narrative went on to say that:

> One must be aware of the growing role of government expenditures as a significant part of the nation's economic activity. For example, in 1975 government accounted for 13.2% of the GNP, while in 1950 government accounted for 23.8 billion of a GNP of 286.2 or only 8.3% of the GNP.

The narrative by itself does not sufficiently emphasize this change. But the addition of a *bar chart,* such as that in Figure 3-3, accomplishes pictorially what words alone cannot achieve.

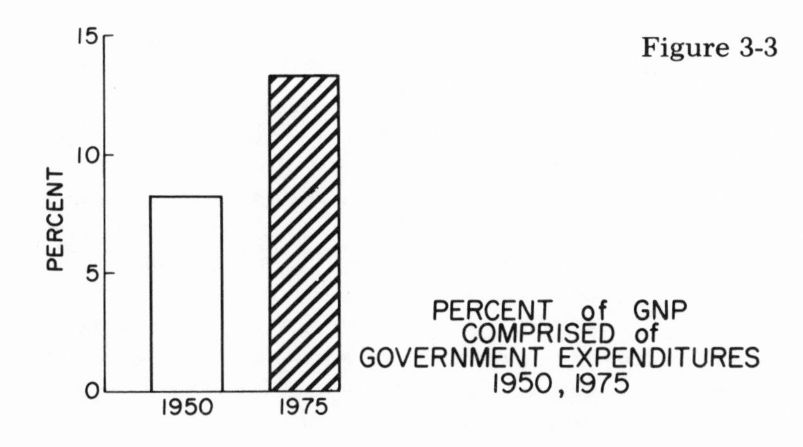

Figure 3-3

SUMMARIZATION

One of the most frequent criticisms of narrative reports is that people must read everything in the report when they are often interested only in the main points. Report writers recognize this problem; therefore, many reports include summaries at the end of various sections of the report as well as a narrative summary or abstract of the total report. Writers like to include as much as possible in their summaries, but, of course, by definition a summary must leave out most of the material.

The use of certain graphic techniques can make it possible to respond to the needs of both writers and readers of reports —that is, the amount of information covered in a summary can be maximized and, at the same time, the quantity of material that the reader must apprehend can be minimized.

The pie chart shown earlier is an example of this device. The chart provides the reader with a summary that covers all the major points in the narrative. One can imagine a much longer narrative, perhaps five or ten pages, that describes the fine points in the breakdown of the GNP. Yet the same pie chart would serve as an adequate summary of such a lengthy description.

Figure 3-4, which is a *frequency distribution graph,* is an example of a way to summarize the following excerpt of the narrative presented at the end of a lengthy section of a report.

> The main question that the foregoing data address is how long it takes to begin to realize maximum performance from our new employees. The study conducted last year showed that, in the aggregate, the average time that productivity was maximized was eighteen months after employment. However, this aggregate average is somewhat misleading for purposes of organizational planning. This is because there were significant differences among different levels of employees. These detailed findings are included in the earlier tables. It should be pointed out here that managerial personnel reached their highest levels of performance after one year of em-

ployment; technical and professional staff after eighteen months; and line workers after two years. But these are not evenly distributed gains among each of the three groups—that is, some groups approach peak performance at an earlier point than others, but do not realize their complete maximum until a later time. For example, technical and professional staff come close to peak performance after twelve months, but actually require six more months to peak.

Figure 3-4

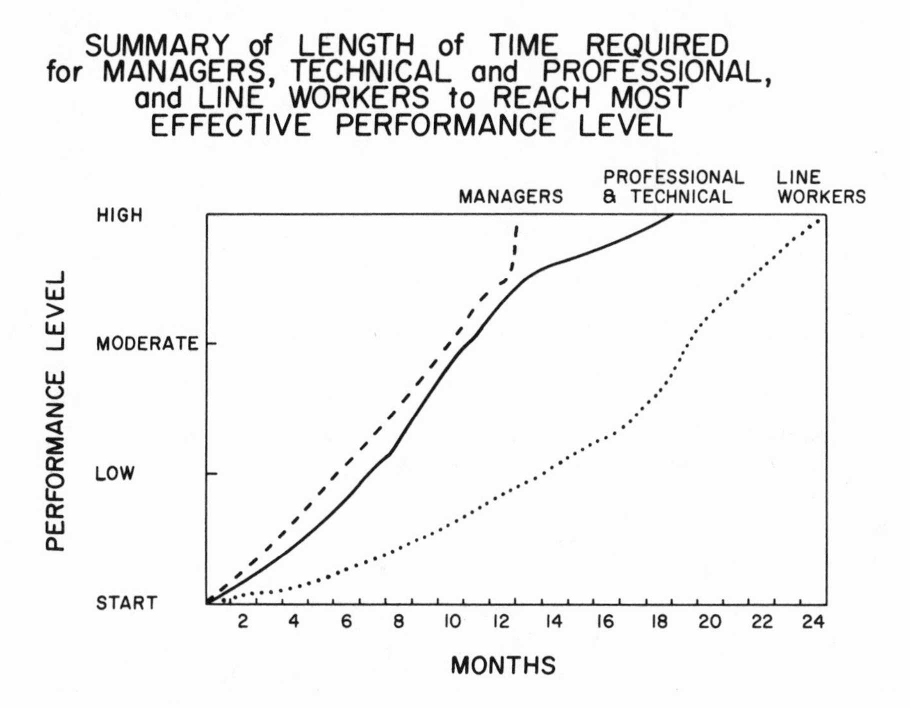

▶ REINFORCEMENT

It is well recognized that one of the principal ways to get people to remember as well as to understand a particular idea is through repetition. There is an old axiom in the advertising business that an effective advertisement must mention the product name at least three times. It is also realized that presentation of the same point in more than one way also contributes to recall and understanding. Different

Figure 3-5

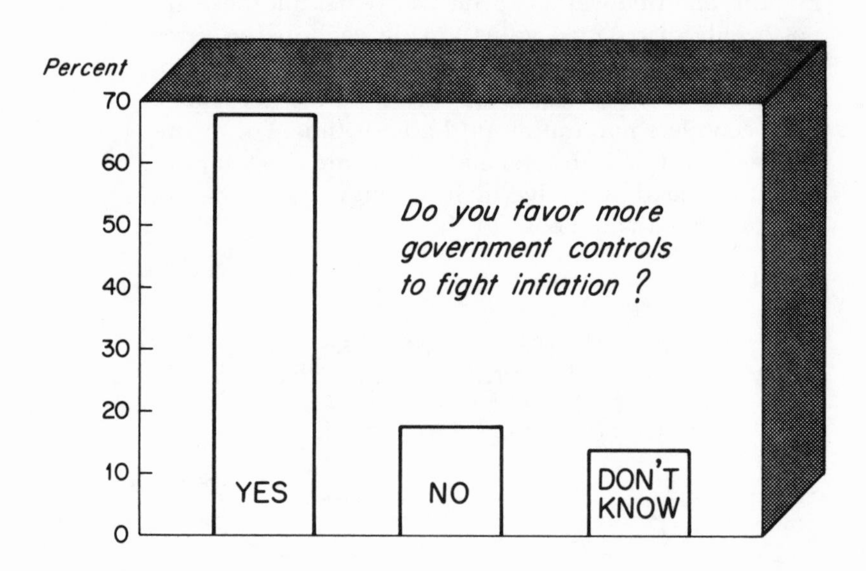

people learn in different ways. Different people perceive and comprehend information in a variety of ways. A reader can completely miss a point made in one form, but may understand the same point when made in another manner.

Writers, therefore, can increase the possibility of retention and understanding by using more than one form to present the same material. Graphics provide writers with an opportunity to represent narrative material in a form other than words and numbers—that is, in a pictorial manner. The use of graphics can serve to reinforce the narrative. This, in turn, contributes to recall and understanding. One must, however, be selective in the process. The writer must decide what material needs reinforcement on the basis of what is most critical for achieving the report's main goal. For example, a report may have as a goal convincing a policymaker not to make a decision in a certain area or, conversely, to make a particular decision. The material selected for graphic presentation for purposes of reinforcement should be that information that is most persuasive in terms of the sought-after action.

The following is an excerpt from a narrative report that is

reinforced by the use of the *bar chart* illustrated in Figure 3-5.

> One of the most important findings from the poll is the declining confidence shown by the respondents in the ability of the free, uncontrolled marketplace to slow down inflation. Only 18% of the respondents think that less government intervention in the economy would help to arrest the rate of inflation, while 68% believe that more strenuous government efforts are needed. Some 14% said they didn't know whether they favored more or less government controls.

▶ INTEREST

Perhaps the most frequent criticism of the kinds of reports with which this book is concerned is that they are "boring." One factor that contributes to this impression is that most reports rely only on the written word. In addition, most report writers are not authors in the sense of being professionals in the use of prose. Further, often the material that one has to present in a report does not lend itself to creating interest, excitement, or drama.

With the ever-increasing use of pictorial material on television, in advertising, in magazines, and in newspapers, readers are more and more conditioned to respond to pictorial presentations. Thus one of the effects of using graphics in reports is to provide an alternative to narrative description that reduces the tedium of words. The periodic insertion of a graphic also serves to "break up" the written text. The reader has to use different perception tools in examining the graphic. When looking at a picture the reader is "resting" some of the usual mechanisms used in reading and comprehending words. Graphics can provide a refreshing pause after which the reader will pick up the narrative with renewed vigor. In this way graphics make the report more interesting, both visually and cognitively. Note the use of a variety of symbols in this book that contribute to making the text more visually interesting. They include use of bullets (●), squares (■), and points (▶) as substitutes for numbers in presenting listed items and in highlighting headings and subheadings.

▶ IMPACT

In addition to providing emphasis to certain points within a report, graphics can contribute to the impact of the total report—that is, readers tend to pay more attention to and remember a report that has graphic presentations. As a result such reports are also more persuasive. Because of the great volume of written reports that are prepared and distributed, writers are almost always competing with other writers' reports. Graphics help get attention.

Graphics can be used to motivate readers to read the report rather than putting it aside or filing it away. One technique to do this is to use an illustration on the cover page or title

Figure 3-6

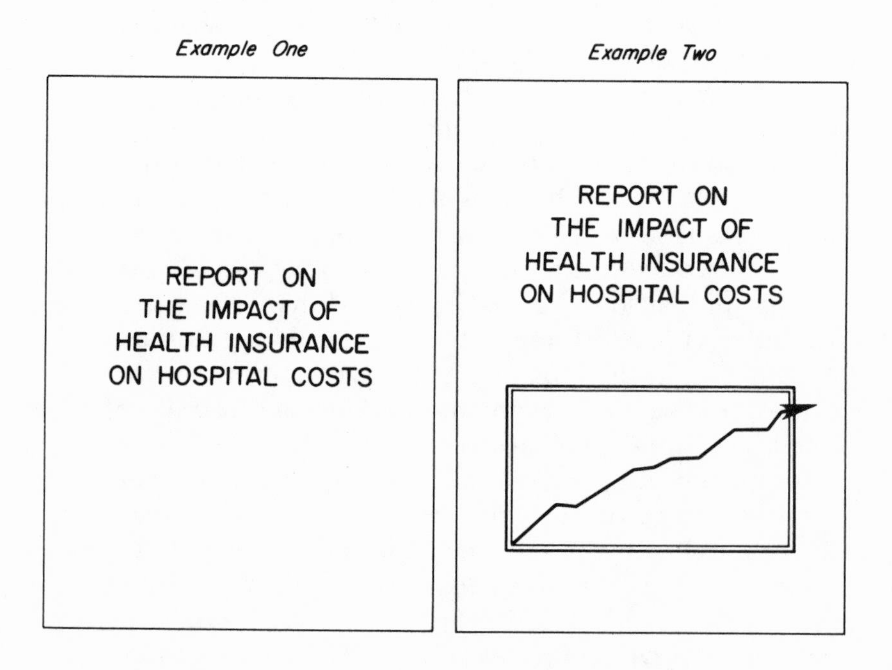

page of the report. For example, take the two sample title pages in Figure 3-6. The second example, through a simple graphic, encourages the reader to turn the title page and "get into" the report.

Note also the use of the symbol ▶ to accompany each of the headings in this chapter. This simple device serves to give interest to the appearance and emphasis to the heading; and it gives the chapter more overall impact than if words alone were used.

▶ CREDIBILITY

Writers want readers to attribute validity to their presentations. They want the reader to believe that the writer knows what he or she is talking about. They want the reader to be impressed and to respect the ideas and the numbers that are presented.

Graphics contribute to the credibility of a report for a number of reasons. Graphics appear to have a certain preciseness to them, and this creates the impression of accuracy of the report's information. Readers also often do not possess skills in graphic techniques and are impressed by the writer's use of these methods of presentation. In addition, because graphics are basically pictures, they give readers the effect of "truth" in what is being presented. That "pictures don't lie" is a generally accepted truism. Modern technology, of course, makes it relatively easy to lie with pictures. Yet the notion that a picture has more authenticity than the written word is still held by many people. Thus readers are impressed by pictorial representations in two ways. First, they believe in their authenticity; and second, they have a certain awe or respect for the ability of the writer to use graphic techniques.

Further, graphics convey a sense of surety or exactness. Readers tend to believe that the writer is sure of his or her material, since graphics seem so precise. There is no hedging; no "yes, but"; no "on the other hand" in a graphic representation.

In view of this fact, writers must be careful not to mislead themselves and their readers when using graphics. If the information presented in a narrative must be considerably qualified, if a number of provisos are necessary, if the meaning must be modified, then a graphic presentation is not appropriate. It is perfectly legitimate to use graphics to impress; but their use to create a false impression should be avoided because in the long run this reduces the writer's credibility.

It is improper to use graphics to attempt to get the reader to believe something that is not necessarily true; to emphasize a point beyond the importance of the substance of the material itself; or to bring readers to a state of confusion through an impressive *form* of presentation that makes them throw up their hands in confusion while still believing that the writer knows what he or she is talking about. Graphics, because of their pictorial quality, can easily be used for such purposes. The multiplicity of lines and forms, all seen at once, can be a more effective and immediate way to mystify than words. You should avoid such mystification if you want to establish authentic credibility.

▶ COHERENCE

One major factor that contributes to the effectiveness of a report is the extent to which the argument being presented comes through in a way that is an understandable, coordinated whole. In an effective report the various points that are covered should hang together. The relationships between or among different points should be made explicit and clear.

Graphics can be used to increase the coherence of a report because they can be used to show:

- the relationships between or among different points
- how the different points are coordinated into a holistic presentation.

In the example that follows, the relationships among the points made in a single paragraph are made more explicit by the use of the graphic *flow chart* that is illustrated in Figure 3-7.

The application for approval of all research proposals must be submitted by June 1, 1981. Applications are then reviewed by the agency's staff to ascertain if they are complete, follow the agency's rules and regulations, and are consistent with and responsive to the legislative mandate that established this research program. Only proposals in the $40,000–$100,000 total request range will be eligible for further review as stated in the program's guidelines. The staff will complete this review by August 1, 1981, and will then transmit the proposals to the appropriate review panel in each of the three fields of support: aging, mental health, and delinquency. The panel members will read and score each proposal according to the criteria previously distributed. These individual reviews are scheduled to be completed by September 15, following which the three panels will meet and review each proposal and the individual scores and will rank proposals. The top ten proposals for each panel will be recommended for funding to the agency. Panels will complete their ranking and recommendations by October 1. The director's office will then notify each applicant of the outcome of the review by December 1.

APPLICATION REVIEW PROCESS Figure 3-7

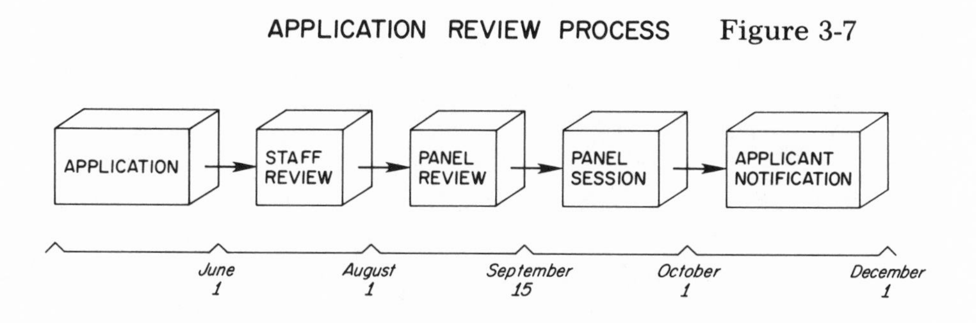

Another example of how a graphic contributes to coherence is illustrated in Figure 3-8. This chart graphically summarizes the material in this chapter regarding how graphics can be used to accomplish one or more of nine related objectives.

Once you have your strategic objectives well in mind you are ready to begin to prepare the graphics for your report as described in subsequent chapters. First, however, you will need a few basic supplies to make the job easier. These are described in the next chapter.

Figure 3-8

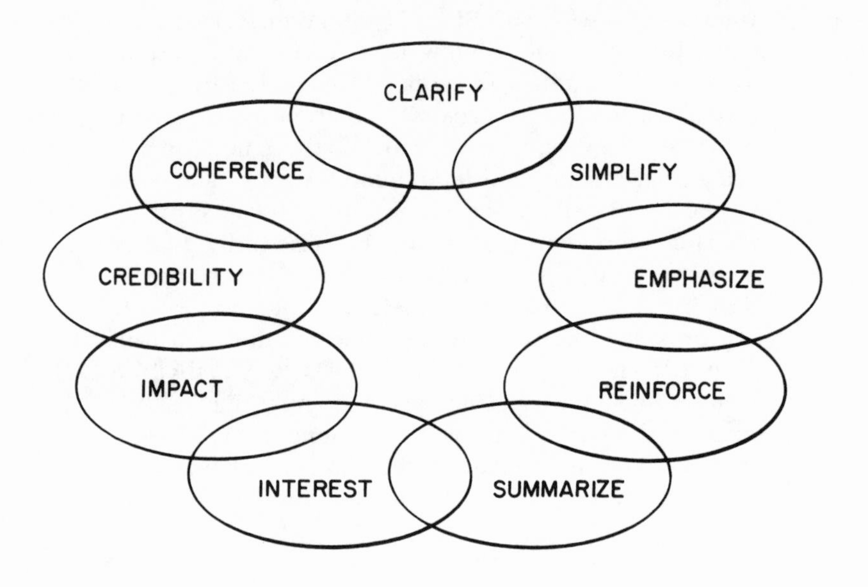

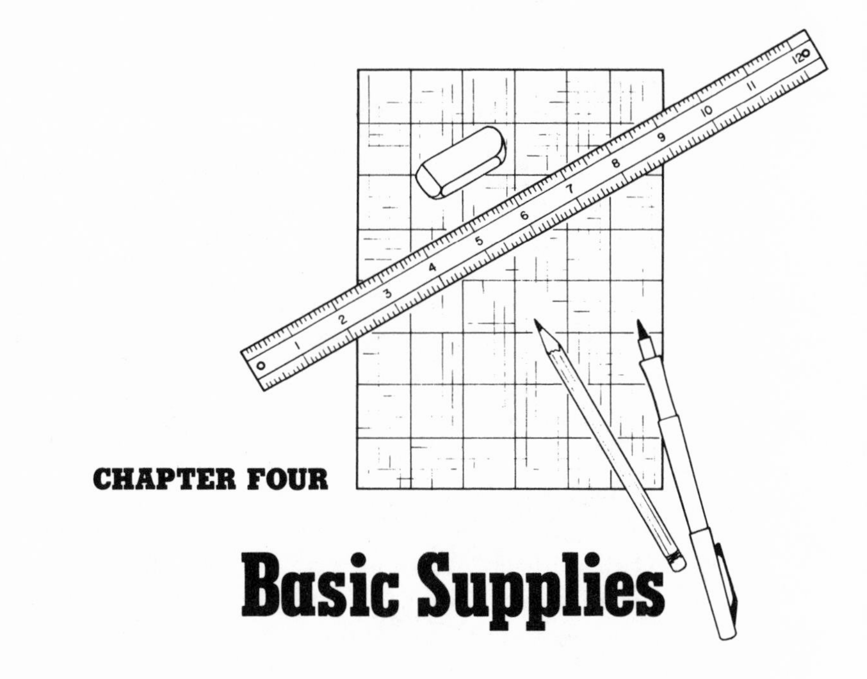

CHAPTER FOUR

Basic Supplies

All the charts and other graphic illustrations that are explained in this book can be prepared with a minimum of special supplies. These include:

- felt-tipped or similar ball-point pens with heavy, medium, and fine points
- graph paper
- No. 2 pencils (medium)
- a gum eraser
- white correction fluid
- a ruler
- plastic templates such as the computer diagrammer, giant circle, plotting mark, and organizational chart templates
- pressure-sensitive shading sheets and charting tape

These materials will be described briefly at this point. They are explained more fully in Chapter 12, "Resources."

• PENS

Use only black pens, since these reproduce the best from copying machines.

The lines of your charts should stand out, and the best thickness for lines can be achieved with medium-point felt, nylon-tipped, or similar pens. Inexpensive pens work fine. Use the heavier tips for the lines of the various shapes (rectangles, circles, etc.) or for plotting lines that comprise the chart. Use a thinner tip for connecting lines, scales, and lettering.

• GRAPH PAPER

Almost all charts should be initially drafted on lined graph paper (also known as coordinate paper, grid paper, and cross-section paper), which is sold in pads or packages in office-supply, art, and stationery supply stores or departments. Graph paper comes in a wide variety of squares to the inch. The easiest ones to use have either four, five, or six lines to the inch, although any size up to ten to the inch will suffice. After you draft your chart on graph paper it can be copied onto regular typing bond by placing the bond over your draft. You will be able to see the heavy black lines of your chart through the bond paper and trace the chart. Guide the recopying with a ruler or template. Be sure the pages are lined up evenly. Some manufacturers make a graph paper that will not reproduce the grid lines, and recopying can sometimes be avoided.

• PENCILS AND ERASER

Preliminary sketches and tracings of charts can be made with No. 2 medium pencils. Use a good eraser such as a gum or ruby eraser that will not leave smudges.

• CORRECTION FLUID

Any lines you do not want on the final copy can be eliminated with white typewriter correction fluid so that they will not reproduce when you use a copying machine. Water-base fluids are the most convenient.

• TEMPLATES

Using plastic templates enables you to prepare drawings that will look as professional as if they were done by a commercial art department. A template is a sheet of transparent plastic with a number of different symbols that are cut out to enable you to trace them.

The single most useful template is a computer diagrammer. It, as well as the other templates described here, are available in office-supply and stationery stores for three or four dollars. The computer diagrammer has a variety of squares, arrows, rectangles, circles, and other shapes and symbols. You don't have to know anything about computers to use it, so don't let the name intimidate you. Its edges are ruled for measuring, and the whole sheet is lined similarly to graph paper so that it can always be lined up with the graph paper to assure that all lines are straight.

The second most useful template is one with circles of different sizes. Use this for pie charts and other charts requiring larger circles.

A third template of plotting symbols is useful to prepare different markings for line charts and to prepare many other symbols to add interest and variety to the appearance of the narrative of the report.

Another useful template is an organization-chart template, which is especially suited for preparing organizational charts and flow charts.

• TRANSFER SHEETS AND CHARTING TAPE

There are numerous kinds of transfer sheets available in office-supply and art stores that can be used to achieve various shadings and toning for charts. There are two types of sheets. One type of sheet is used by rubbing it with a stylus so that the shading comes off on the paper underneath. The second kind, which is preferable, is self-adhering. It is used by cutting out pieces of the desired shape and size, removing the backing, and pressing it on the chart. The most satisfactory way to cut out the pieces is to place the sheet over the chart and lightly cut it to shape with an Exacto No. 1 knife.

Self-adhering charting tape comes in rolls of various widths in solid black or in patterns. Just roll out the length you need and press it on the chart.

• SOME OPTIONAL SUPPLIES

If you have the foregoing supplies, you don't really need anything else to prepare the graphics illustrated in this book. But there are some additional optional supplies that you might find useful as you develop your own style of preparing graphics. These include a compass, a protractor, and rubber cement.

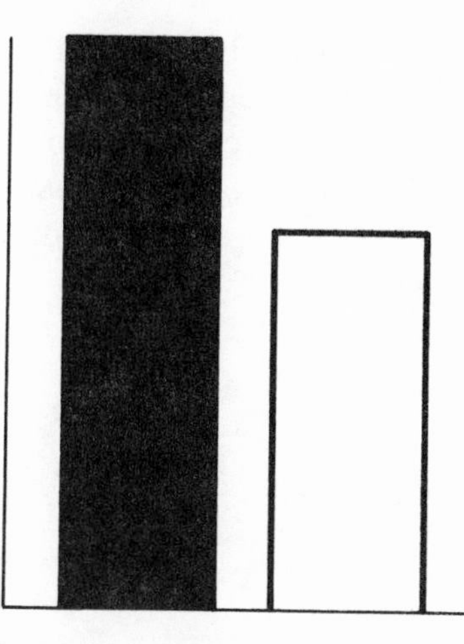

CHAPTER FIVE

Bar and Column Charts

WHAT IS A BAR CHART?

The bar or column chart is the easiest type of graphic to prepare and use in reports. It employs a simple form: four straight lines that are joined to construct a rectangle or oblong box. When the box is shown horizontally it is called a bar; when it is shown vertically it is called a column. Unless otherwise noted, we shall use the term bar chart to refer to both bar and column charts.

Bar charts are most useful for illustrating quantitative information to be emphasized in the report. The bar chart is an effective way to show comparisons between or among two or more items. It has the added advantage of being easily understood by readers who have little or no background in statistics and who are not accustomed to reading complex tables or charts.

Every bar chart is comprised of two or more items of information, each of which is represented by a bar. For example, suppose a report makes reference to the fact that there is a difference between the participation of men and women in a program: men comprise 30% and women comprise 70% of the participants. If you want to impress this comparison on the reader, you can supplement your text with a vertical chart such as that shown in Figure 5-1 or a horizontal chart as in Figure 5-2.

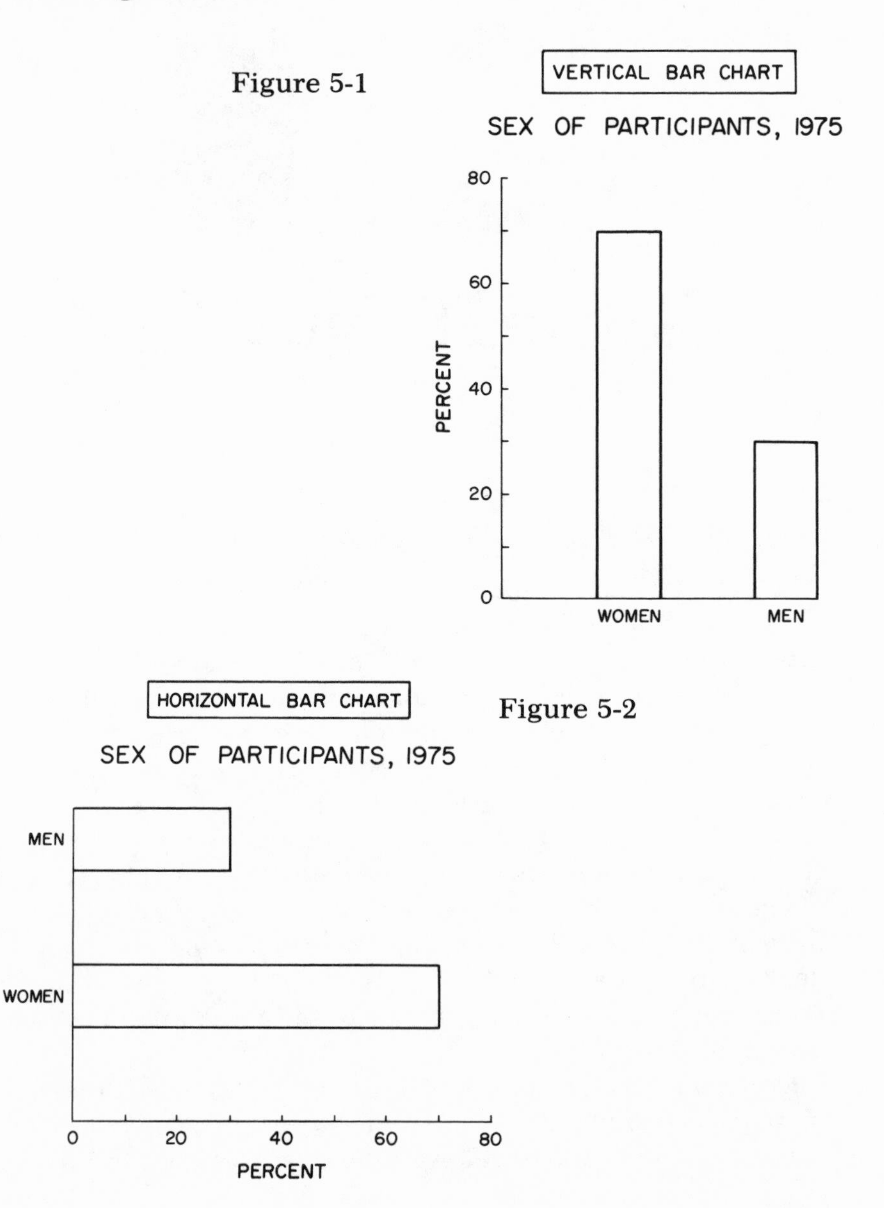

Figure 5-1

Figure 5-2

WHAT DO BAR CHARTS ACHIEVE?

Bar charts can be used most effectively to achieve:

- clarity
- simplification
- reinforcement
- interest

Bar charts can be used to show percents as in Figures 5-1 and 5-2, or to show absolute numbers as in Figure 5-3.

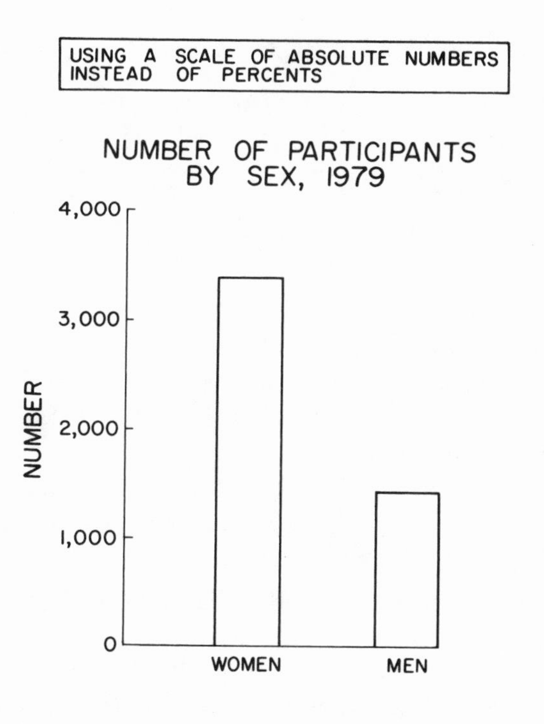

Figure 5-3

BAR OR COLUMN?

The writer must decide on the direction of the bar. Some believe that the vertical bar should be used when comparing similar items for different time periods and the horizontal bar for comparing different items for the same time period.

However, most people find the vertical-bar format easier to prepare and read, and a more effective way to show most types of comparisons. You can judge this for yourself by deciding whether Figure 5-1 or Figure 5-2 is more effective.

A vertical chart is usually easier to fit on the standard 8½ ×11 paper used in reports, because when the bars run vertically, there is more room on the page for the longest bar. This format reduces the need to break up the longest bar and also makes it unnecessary to reduce the size of the whole chart to make it fit the page. The vertical chart is also easier to label because there is more room on the page to type in the labels.

SHOULD THE BARS BE SHADED?

Bars can be drawn using only lines, as in the earlier examples. The white of the page gives the bar its body. The bars can also be shaded or hatched in order to add:

- interest
- emphasis

Shading and hatching of the bars contribute to the differentiation of the factors being compared. Figure 5-4 is based on the same information as the earlier charts, but is shown using hatching for one of the bars. Notice how this makes the chart more interesting to look at and how it sharpens the comparison between the two factors being shown. In Figure 5-5 we see how four different bars are each given a different shading, with self-adhering shading sheets used to accomplish the effect.

Shading of this type can also be done by hand, but the use of prepared shading sheets or charting tape is generally more effective. These materials can be bought in most good art, stationery, or office-supply stores under a number of brand names that are listed in Chapter 12, "Resources."

There are three main categories of shadings: hatching and cross-hatching, a range of toning shades, and dots of different size and intensity. The material comes in a variety of kinds and makes of plastic-like sheets. Contact sheets are

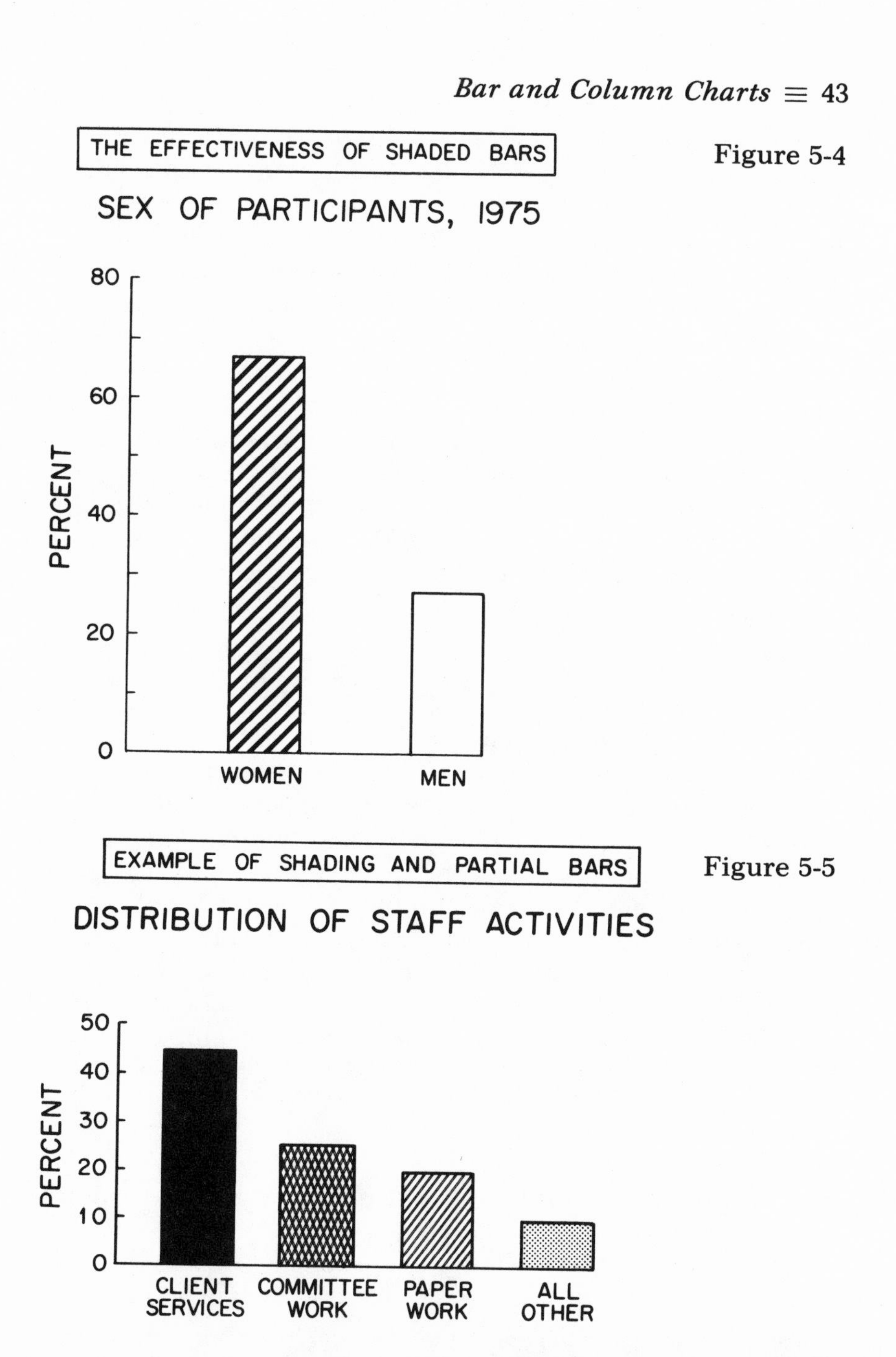

Figure 5-4

Figure 5-5

placed over the bar and rubbed with a stylus or pencil, and the shading comes off on the bar. These sheets can be messy to use and often hard to line up with the bar. Self-adhering sheets are generally more satisfactory. Using a razor or sharp knife, one cuts a piece of the sheet to fit each bar, removes the backing, and sticks it on. A wide variety of patterns and shadings are available in this material. A very satisfactory alternative is to use pressure-sensitive charting tape, which comes in rolls of various widths in solid black and in as many as ten different patterns. You just cut it to the length you need and stick it on like transparent adhesive tape.

When using shading with horizontal bars, it is advisable to use the darkest shades on the bars closest to the bottom of the page. With vertical bars the darkest shading should be used for the bars on the side of the page. Make the longest bars the darkest and reduce the intensity of the shading as the bars get shorter, as in Figure 5-5.

PARTIAL BARS OR WHOLE BARS?

Up to this point all bar charts that have been illustrated employ the use of partial bars—that is, each bar in the chart shows what part it is of some larger total, such as 100% or 10,000 people. This is an effective way to compare the relative size of one factor to one or more other factors, since each is represented by a single partial bar.

It is also possible to use a single bar and divide it into its component segments, as in Figure 5-6. This method has the added advantage of enabling you to show the relationship of each factor to the whole. Figure 5-6 shows the same data as Figure 5-5. As you can see in Figure 5-5, using partial bars is

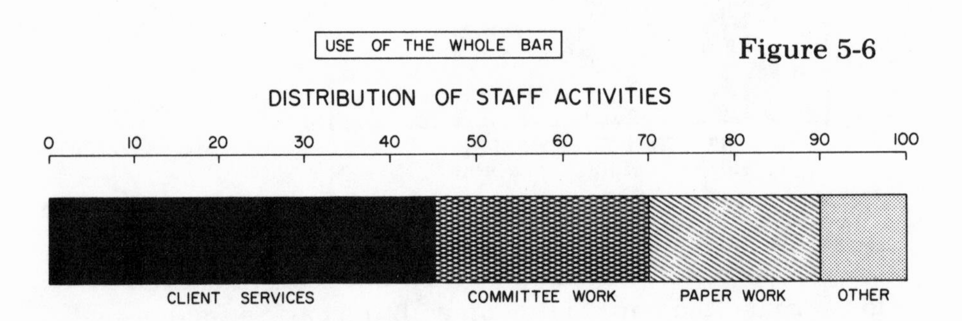

Figure 5-6

a more effective way to compare the relationships among each of the four factors shown. Figure 5-6, on the other hand, is more effective in showing what proportion each of the four factors is of the whole 100%.

Bar charts using whole bars can also be prepared using more than one whole bar, as in Figure 5-7. This technique allows you to show comparisons of the relative size of two or more factors (for example, men and women) over time (for example, 1970, 1978) as well as to show the proportion each is of the whole (100%). Thus the use of whole bars assisted by shading is a way to increase the amount of information conveyed in the chart.

Figure 5-7

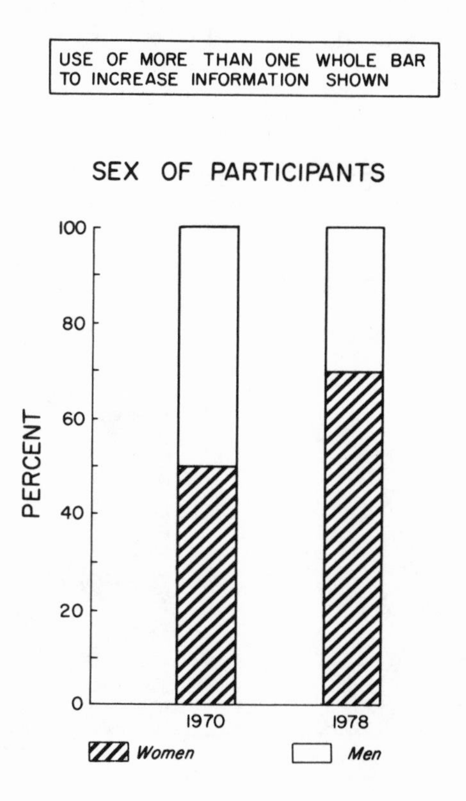

When different shadings are used within the same bar, the darker shadings should be at the beginning of the bar—that is, at the bottom of a vertical bar or on the side of a horizontal bar that abuts the axis showing the unit of analysis (usually the left side).

INDEPENDENT OR JOINED BARS?

The bars may each stand independently with a space between each bar, as in the prior examples in this chapter. Or the bars may abut each other and be joined together as a group. This type of graphic is illustrated in Figure 5-8. It should be noted that when bars are joined together, less space is available for the labeling. Since labeling in reports is often done by typewriter, this can create some problems in having sufficient lettering space. Therefore, it is generally easier to leave space between the bars. It is also somewhat easier to read spaced bar charts than those in which the bars are joined. In addition, the comparison or contrast between the different bars tends to stand out more in a chart where the bars are spaced.

USE OF JOINED BARS — Figure 5-8

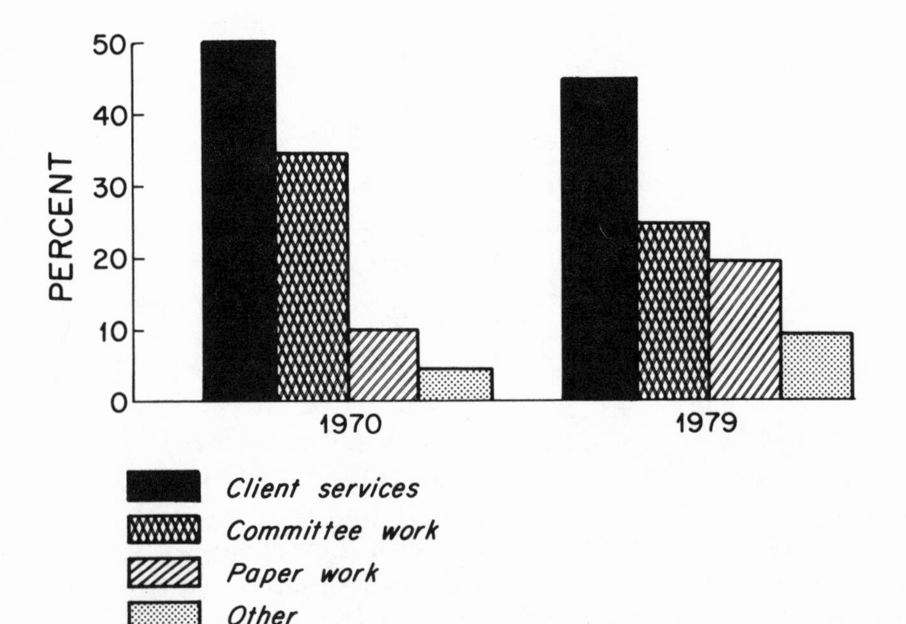

Nevertheless, joined bar charts do have some important advantages. One is that if your chart has a number of bars (ten or more), joining the bars may be the only way to fit it on the page without making the bars too narrow. Another major advantage is that the joined bars can be a very effective way of *comparing groups* of similar information for two or more different time periods or by some other factor. This is shown in Figure 5-8, in which data for two different time periods are grouped together. When using two or more groupings of joined bars, you should try to keep the number of joined bars in each group under four. In addition, it is always helpful to shade the bars within each group as in Figure 5-8. This provides the contrast needed for the reader to make the distinction between each joined bar.

The distance between each group of bars should be approximately one fourth or more of the width of each group and should be at least as wide as any single bar.

USING SHADING AND A LEGEND

Shading the bars can be done simply for visual purposes or can add additional information, as in Figures 5-7, 5-8, and 5-9. In effect the shading adds an additional unit of analysis. When bars are shaded for this purpose, it is necessary to accompany the chart with a legend or key that explains what the different shadings mean, as in these figures. Note how much information is conveyed in these charts by using a legend, and shading and joining the bars.

The selection of the legend is important since the legend will determine what aspect of the information is emphasized. For example, in Figure 5-9 the same information is presented as in Figure 5-8, but a different legend is used. Both charts show a decrease in client services and an increase in paper-work activities. However, while Figure 5-8 emphasizes the changes in the pattern of activities from one year to another, Figure 5-9 gives more emphasis to the changes in each activity. The reason for the different emphasis in the two charts is that the main point emphasized is controlled by the selection of the unit of analysis used on the bottom or

Figure 5-9

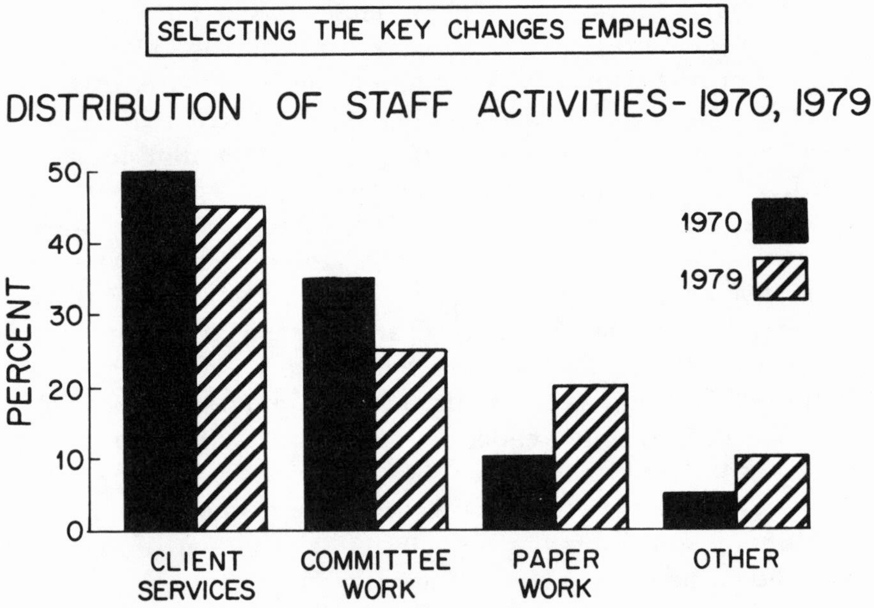

SELECTING THE KEY CHANGES EMPHASIS

DISTRIBUTION OF STAFF ACTIVITIES- 1970, 1979

horizontal axis (years in Figure 5-8, activities in Figure 5-9). The selection of the different unit of analysis, of course, results in a reversal of the legend—activities in Figure 5-8, years in Figure 5-9.

LABELING AND LETTERING

In order to be sure that a bar chart accomplishes its purposes of simplifying, clarifying, and emphasizing your material, it is important that it be properly labeled. This can be done by typewriter, by hand, or by using pressure-sensitive lettering available in art and office-supply stores (as explained in Chapter 12, "Resources").

Labeling the chart includes:

- A title for the chart that describes what the chart shows, including reference to the unit of measurement, unit of analysis, year or years for which the chart is relevant, and the place to which it refers.
- Showing the unit of count or value that is being used, such as percent, number, or rate along one axis of the chart.

- Describing the information that is being shown, such as "men," "women," "black," "white," "1975," along the opposite axis.

Labeling of the information can be done adjacent to the bars or can be done within the bars. Some examples of different labeling options are shown in Figure 5-10.

Figure 5-10

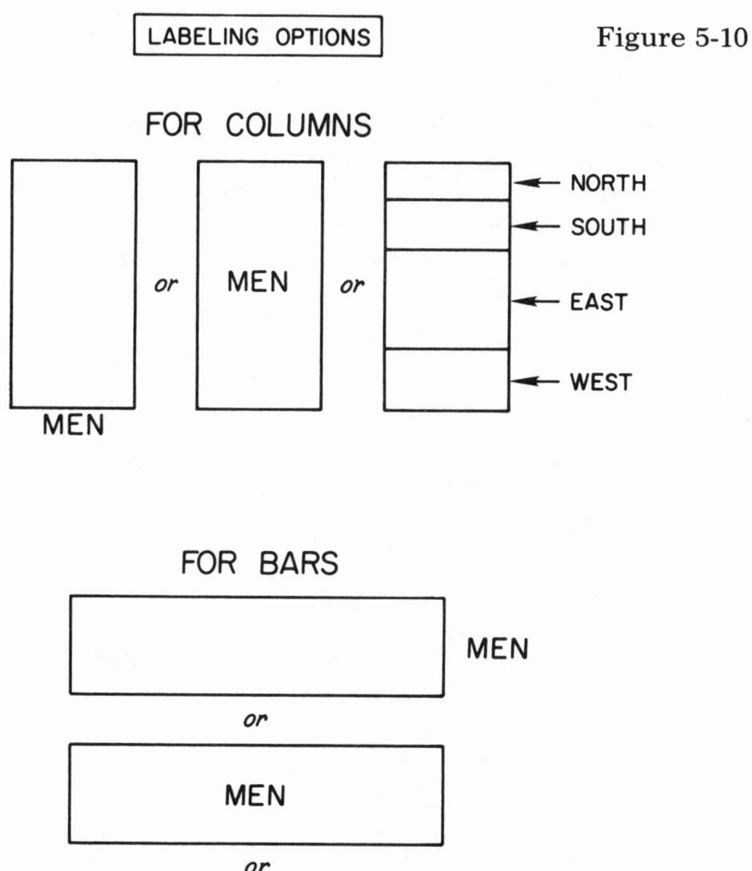

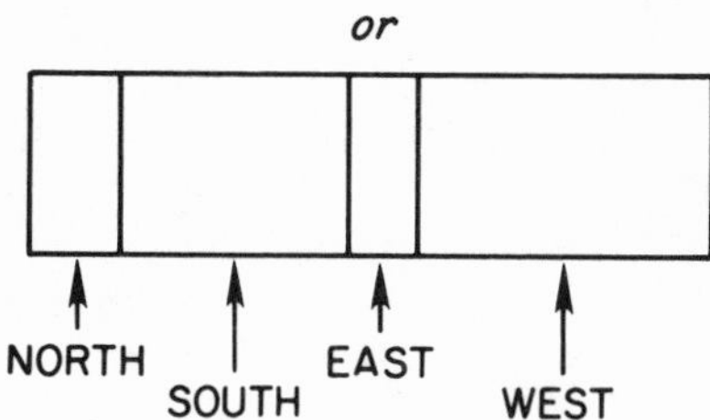

HOW TO PREPARE A BAR CHART

Bar charts are illustrations that show numbers in an easily understood fashion. Because they are drawings, they cannot graphically show the numerical information in a 100% precisely accurate manner. Nor can they show all the detailed information that might be included in the narrative or in a table. Nevertheless, you should try to have the bar chart show as accurate a depiction of the quantitative information as possible. This requires that (a) the length of the bars be related to a uniform scale within the chart, and (b) the relationship of one bar to another be in proper proportion. You can use a number of techniques that help you prepare the bars in proper proportion.

There are seven basic steps in preparing a bar chart. Figure 5-11 shows the components of a bar chart. These are referred to in the following steps:

Figure 5-11

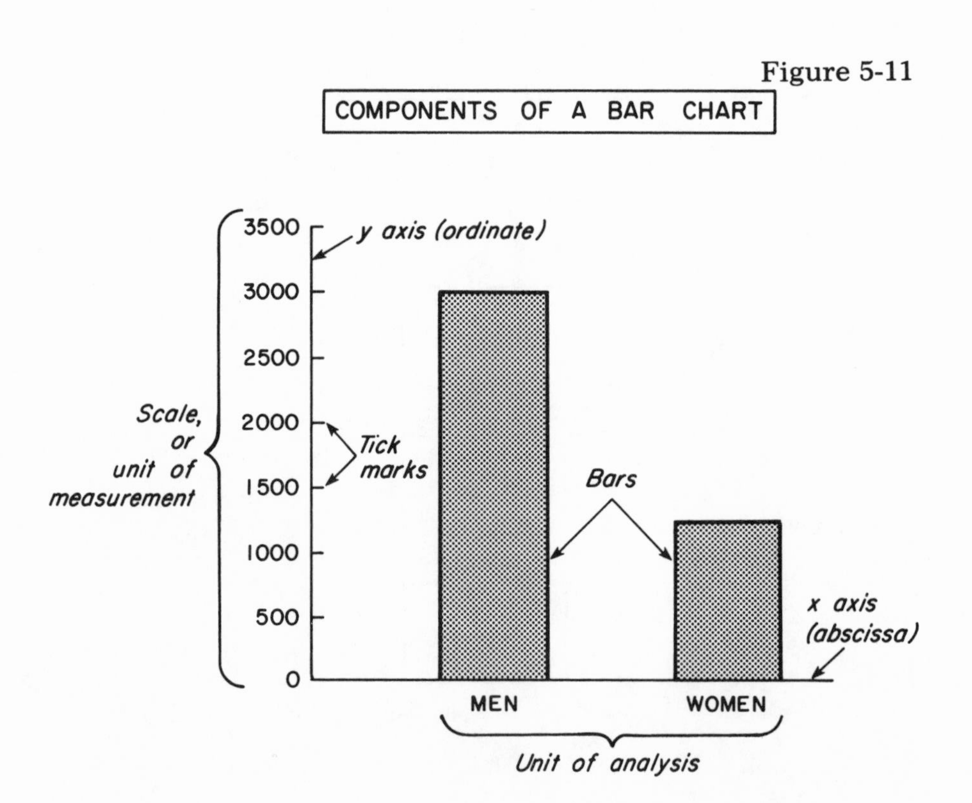

STEP ONE: Planning the Page Location and Size

The first step is to decide how much of the page you want the chart to occupy and the approximate size of the bars. The space can range from about a quarter of the page to a whole page. Be sure to use sufficient space. A frequent mistake in using graphics is to crowd them and thereby lose their impact.

STEP TWO: Decide on the Length and Width of the Bars

Whenever you use a bar chart be sure that every bar in the same chart is the same width. How wide should a bar be? The answer depends on how many bars are in the same chart. Assuming standard 8½″×11″ paper, the following table provides guidelines for the width of the bars.

NUMBER OF BARS IN CHART	WIDTH OF BARS
2	2″–½″
3–5	1″–½″
6–10	½″–¼″

If vertical, the length of the longest bar should never exceed seven inches. If horizontal, the longest bar should not exceed five inches. The spacing between each bar should be sufficient to allow the contrast between the different bars to show up clearly. This usually means that the space between bars should be no less than one half the width of the bar itself. If you have sufficient space to spread the bars out, allow the space between bars to be the same or a little more than the width of the bar. (The foregoing specifications refer to the spacing of independent bars. Spacing rules for joined bars were explained in the earlier section on independent or joined bars.)

Keep the length and width in proportion by balancing the length and width. Plan the dimensions of your chart so that except in very unusual circumstances no bar is ever wider than it is long.

Step Three: Decide on the Scale (Units of Measurement)

The third step is to devise your scale. This is done by defining the units of measurement: numbers, percents, or rates. Then picture your total chart as based on two lines: a vertical line (called the y axis, or ordinate) and a horizontal line (called the x axis, or abscissa). See Figure 5-11.

Step Four: Draw in the Axis and the Scale

Draw the vertical (y) and horizontal (x) axis. Then enter the units of measurement—that is, the scales—on the appropriate axes. The unit of measurement goes on the vertical y axis if the bars are vertical or on the horizontal x axis if the bars run horizontally. (The opposite axis will contain the categories represented by the bars: men, women, age groups, types of activities, years, or any other grouping.)

Whatever unit you select, it is essential to keep the steps uniform and mark them with a tick line. For example, if you want to show numbers of people in units of 500, every point in the scale must go up in units of 500, as in Figure 5-11. Further, the distance must be uniform—that is, each step of 500 must be the same distance from the next step (such as ½ inch). To achieve this, place a ruler along the axis that will be used for the scale, measure the necessary equidistant points, and draw a small tick line coming out of axis. Then print or type the unit's designation (500, 1000, etc., or, 10%, 20%, etc.) next to the axis line. Label the axis "percent," "number," or "rate," depending on what unit you select. This label may run vertically along the side of the vertical axis; or it may run horizontally and be placed just above the top of the vertical axis. (If you run labels vertically, you must of course turn the page sideways in the typewriter.)

Step Five: Enter the Unit of Analysis

The fourth step is to identify the unit of analysis: "men," "women," "black," "white," "other," etc. This will be labeled on the axis opposite from the axis containing the scale after the bars are drawn in. In the usual vertical bar chart this label will be at the bottom of the chart.

STEP SIX: Locate the Bars

The sixth step is to locate the bars. This is done by first locating the end points of each bar. The end point is the point along the scale where the bar will end. There are various ways to locate an end point. One is to estimate it with your eye. Another is to measure with a ruler. Yet another is to draft the chart on graph paper, which is already uniformly lined so that you can select one or more lines to show uniform units of measurement. If you do your chart on graph paper, then you will have to trace or recopy it onto the regular unlined typing paper you are using for your report. Using a transparent plastic template and drafting the chart on graph paper is the easiest and most accurate way to locate and draw in bars.

STEP SEVEN: Draw in the Bars

The seventh step is to draw in the bars. With a little practice, most people do not have much trouble preparing bars that are sufficiently accurate. The biggest problem is to keep the bars straight and in a parallel line with the edges of the paper. If you use graph paper, you just have to follow the lines of the paper with a ruler or side of the template to accomplish this.

If you don't draft the chart on graph paper, there are some tricks you can use to draw bars accurately. One is to measure the distance of the two end points of the bar from the edge of the paper and, being sure they are the same distance, place a dot there. Then draw a line connecting the two dots. Another method is to use a piece of 8½″×11″ graph paper. Place one edge of the graph paper where you want your bar line to be on the plain white paper. Part of the graph paper will then be off the page on which you are going to draw the graph. Be sure that the line of the graph paper that is closest to the edge of your writing paper is straight along the edge of the writing paper. Then draw your bar line by following the edge of the graph paper that is on your plain white paper. You now have one side of the bar. Complete the bar by drawing in the other side and the top line. Simply measure, from the initial bar line, the width of the bar (¼″, ½″, etc.) and draw a parallel line; then join the two parallel lines with a line across the end.

POSITIVE AND NEGATIVE VALUES

The preceding examples of bar charts have all been concerned with showing positive (+) values. Sometimes the material you want to illustrate includes negative (−) values. You can show both positive and negative values on a bar chart by using a scale that includes both. You can also use

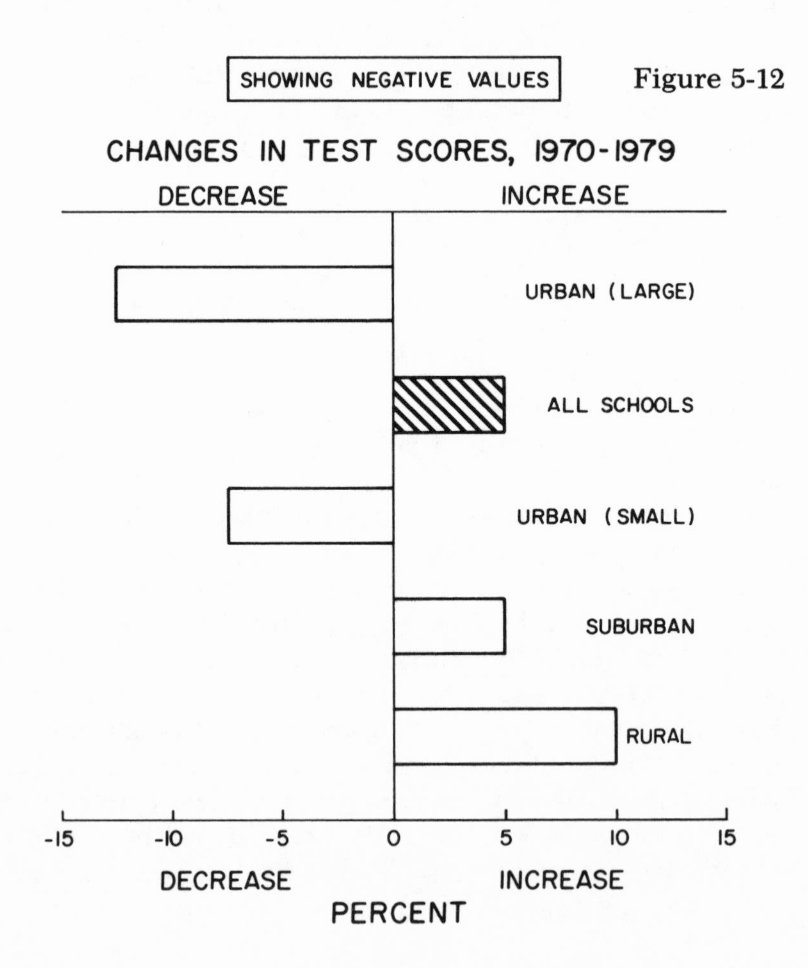

Figure 5-12

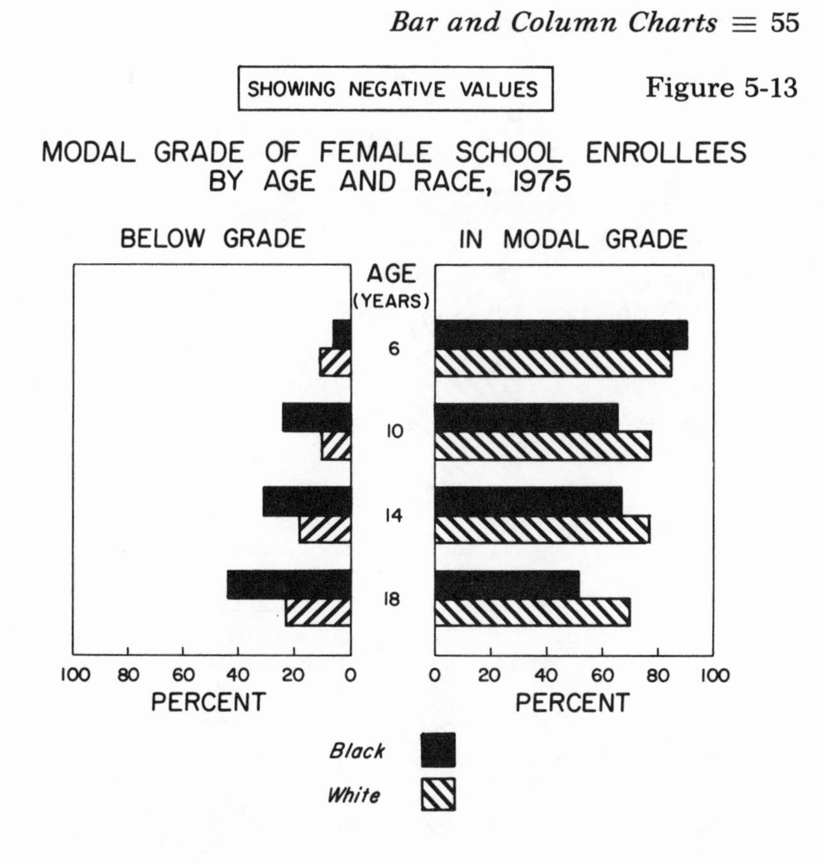

this same format to show increases and decreases, as in Figures 5-12 and 5-13.

SHOWING MORE NEARLY EXACT VALUES

The bar chart, as such, does not show the *exact* quantities that you may have referred to in your text because the lines in the chart are not precisely scaled. If it is important to show the exact values in the chart, this can be accomplished by labeling the exact value either right in the bar, at the end of the bar, or alongside the label. If bars are shaded, it is easiest to put the value at the end of the bar rather than within the bar. Additional preciseness can also be added to a bar by drawing the measurement scale that you used across the entire chart rather than just showing tick marks. These methods are illustrated in Figure 5-14.

Figure 5-14

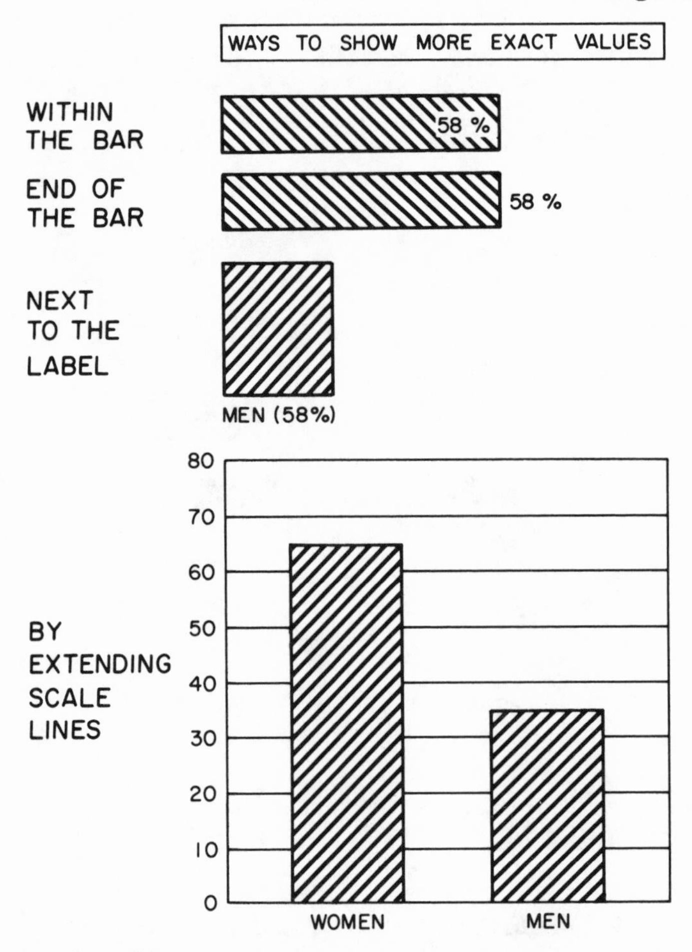

CREATING INTEREST WITH ADDITIONAL DIMENSION

Bar charts can have enhanced impact and can add aesthetic interest to a report if they are prepared so that they appear to have additional dimension—that is, the chart will seem to have depth, body, and volume rather than just be flat on the page. Perhaps the easiest way to add dimension to the bar is just to draw a heavier line at the top and down one side, as in Figure 5-15. A variety of other techniques that involve drawing the bar as a three-dimensional box are shown in the examples in Figure 5-15. Bars of this type can be shaded by using pressure-sensitive tapes, self-adhering shading sheets, by hand, or by using typewriter dots. Bars done in this way can be either independent or joined. A further extension of this method is to box in the entire chart; the boxing can also be shaded. But don't overdo this technique or you will detract from the drawing.

Figure 5-15

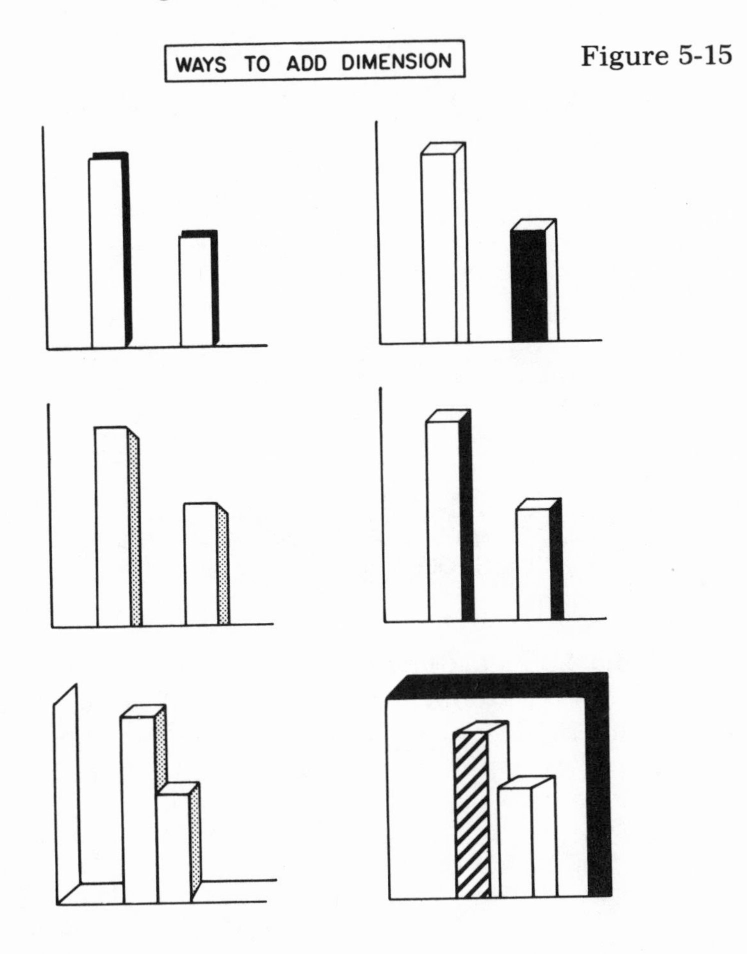

SYMBOLIC BARS

One of the most effective techniques to provide additional interest to a report is the use of symbolic bars rather than rectangular bars. Symbolic bars are comprised of figures instead of lines. These figures can be in the form of people, objects, dollar signs, or other appropriate symbols. They can be purchased in transfer sheets, can be drawn by hand, or can be created on the typewriter.

The typewriter symbols can be done as follows:

```
$$$$$$$$$$$$$$$$$$$$$$$$$$$$$$$$$
$$$$$$$$$$$$$$$$$

QQQQQQQQQQQQQQQQQQQQQQQQQQQQQQQQQ
XXXXXXXXXXXXXXXXXXXXXXXXXXXXXXXXX
QQQQQQQQQQQQQQQQQ
XXXXXXXXXXXXXXXXX

xxxxxxxxxxxxxxxxxxxxxxxxxxxxxxxxx
xxxxxxxxxxxxxxxxx

OOOOOOOOOOOOOOOOOOOOOOOOOOOOOOOOO
mmmmmmmmmmmmmmmmmmmmmmmmmmmmmmmmm
AAAAAAAAAAAAAAAAAAAAAAAAAAAAAAAAA
OOOOOOOOOOOOOOOOO
mmmmmmmmmmmmmmmmm
AAAAAAAAAAAAAAAAA

QQQQQQQQQQQQQQQQQQQQQQQQQQQQQQQQQ
MMMMMMMMMMMMMMMMMMMMMMMMMMMMMMMMM
QQQQQQQQQQQQQQQQQ
MMMMMMMMMMMMMMMMM
```

Symbolic charts are often shown without use of a clearly established axis. They may thus appear to "float" on the page and need to be "anchored." This can be done either by putting in an axis or putting a box around the chart.

HISTOGRAM

The histogram is a special type of bar chart used for presenting quantitative data known as *frequency distributions,* which show the relationship between two factors or variables. It portrays the same data as in the line chart known as a *frequency polygon* or *frequency distribution chart,* which is described in Chapter 7, "The Line Chart."

All the bars in a histogram must be joined. The intervals

Figure 5-16

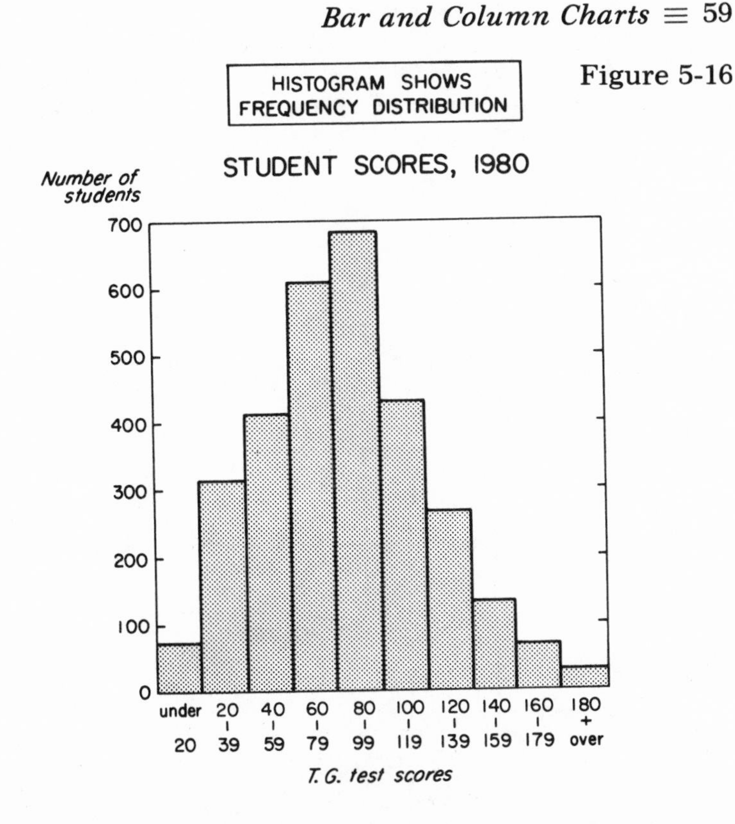

on the measurement scale may be equal or unequal. (However, the histogram with unequal scale distances requires considerable technical skill and is not discussed here.) Histograms are typically used to present epidemiological data, population figures, and test scores where the data are continuous. Each bar in a histogram actually represents a range of scores, ages, or other continuous linear measures known as interval data. In a regular bar chart each bar represents a discrete factor such as sex, race, or occupational groups. A typical histogram is shown in Figure 5-16.

CUMULATIVE BAR CHARTS

Occasionally you may wish to show data that are cumulative —that is you want to show the aggregate amount of some factor as it changes over time. Usually you would use a line chart (known as a *cumulative frequency chart* or a *step chart*), as explained in Chapter 7. However, the bar chart can also be used for this purpose and is exemplified in Figure 5-17. Each bar presents the running total of some value (dollars in Figure 5-17) up to the point in time represented by the bar.

Figure 5-17

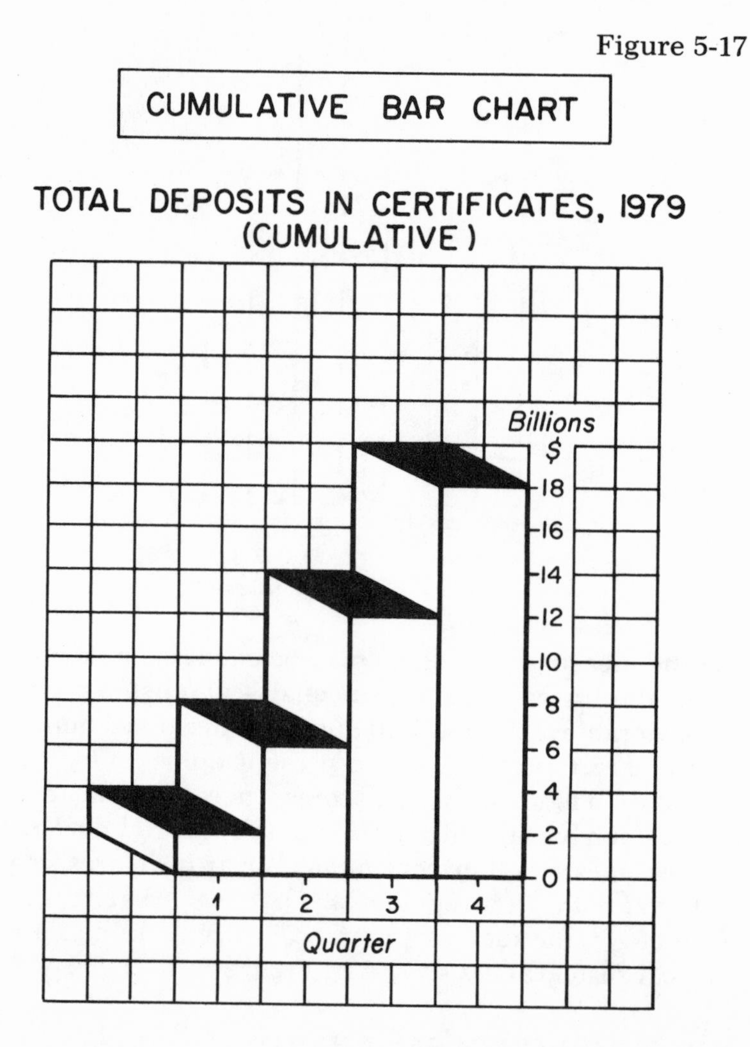

This chapter has covered the basic forms of the bar chart. In the next chapter we shall examine another frequently used chart to display quantitative information, the pie chart.

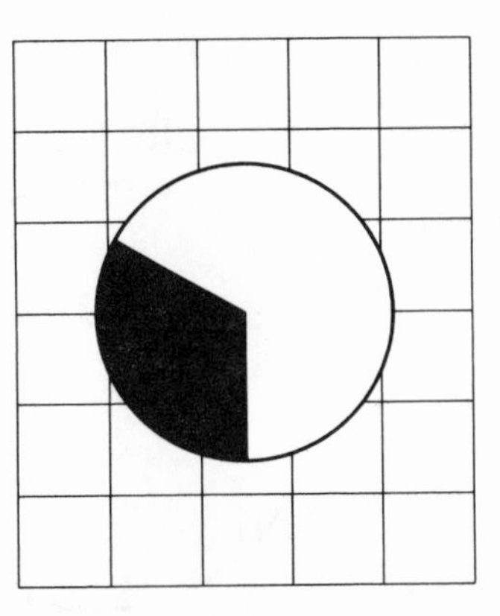

CHAPTER SIX

Pie Charts

WHAT IS A PIE CHART?

A pie chart is comprised of a circle that is divided into segments by straight lines within the circle. The circle represents the total or whole amount. Each segment or wedge of the circle represents the proportion that a particular factor is of the total or whole amount. Thus, a pie chart in its entirety always represents whole amounts of either 100% or a total absolute number, such as 100 cents or 5,000 people. All of the segments of the pie when taken together (that is, in the aggregate) must add up to the total.

Pie charts are also known as "circle charts," "circle graphs," "sectograms," and "sector charts." They are most commonly referred to as pie charts because they portray the size of the pie and the size of each "piece of the pie." Figure 6-1 shows the simplest form of pie chart, one comprised of only two segments.

The pie chart has two basic graphic components:

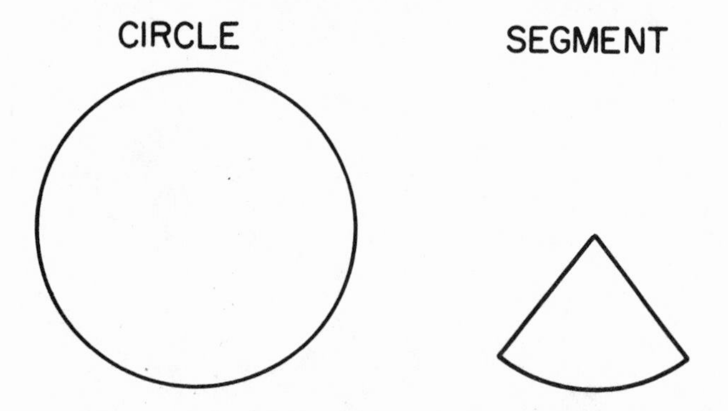

Pie charts, like bar charts, are used to represent quantitative information in either percents or in absolute numbers. They are very popular for showing financial information, such as various sources of income or different areas of expenditures for organizations, businesses, and government. They serve equally well to depict proportions or percents of factors such as population characteristics (men, women, black, white) or any other general categories.

A SIMPLE PIE CHART

Figure 6-1

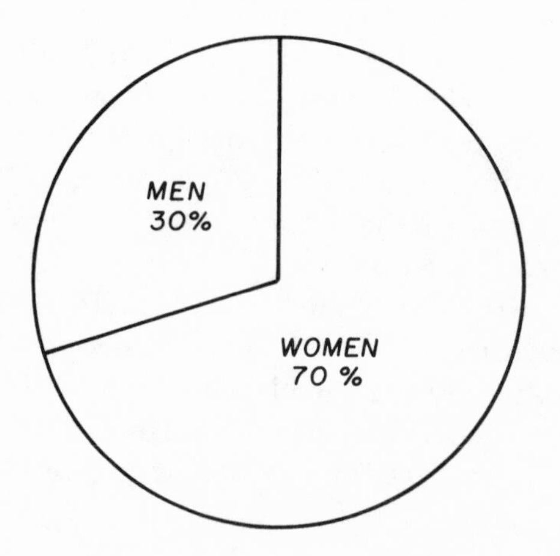

WHAT DO PIE CHARTS ACHIEVE?

Pie charts are most effective as a way to compare the relative proportions of various factors to each other and to the whole. As such, they can be used in a report to achieve:

- simplification
- interest
- summarization
- coherence

The pie chart is a clear and interesting way to present data for which relative proportions are important. The pie chart is most useful in reports intended to help the reader understand and remember proportions and general relationships. However, the pie chart is less precise than some other forms such as the bar chart and the line chart because the pie chart does not have as clearly depicted scale of measurement as do these other charts.

BAR OR PIE?

Both bar charts and pie charts show quantitative information and enable comparisons between or among the values of two or more factors. How can you decide, then, which type of chart to use in your report? In Figure 6-2 we see the same information portrayed in a pie chart, a partial bar chart, and a whole bar chart. By comparing the three forms in this example we can derive some principles to help answer the question.

First, the partial bars are most effective in showing the actual percent of participants—that is, 70% women, 30% men. The whole bar also shows this, but somewhat less effectively. The pie is least effective in showing the actual percents.

Second, the pie is most effective in showing that women comprise by far the larger proportion of the total of *all* participants. The whole bar also shows this, but is not so effective from a pictorial standpoint. The eye/mind tends to perceive circles as wholes more than the oblong box that comprises a bar. On the other hand, the whole bar, because

Figure 6-2

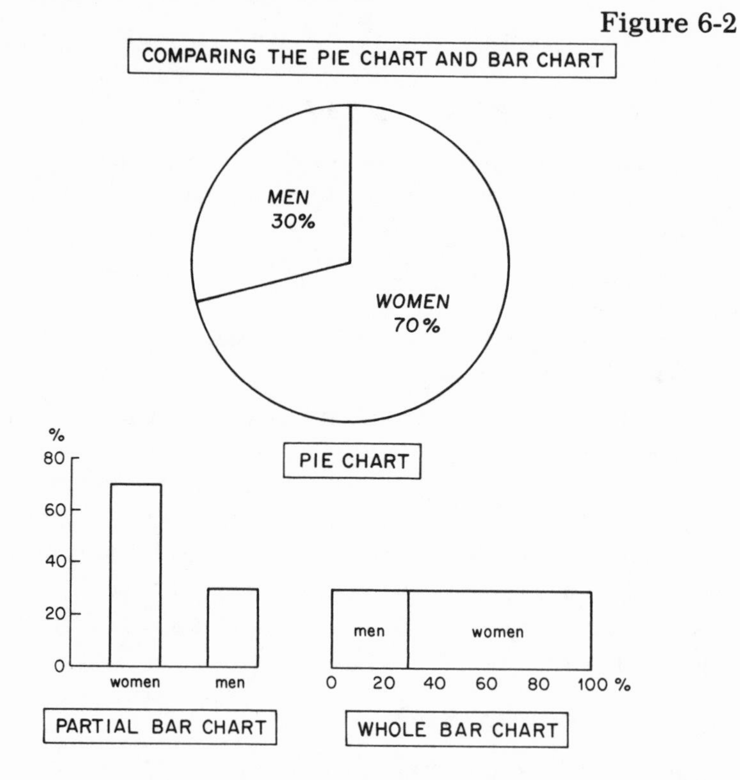

it includes a measurement scale, helps the reader to understand the actual percents—that is, 70% and 30%.

Thus the answer to the question depends on the major effect you aim to convey in the report. If you want to dramatize comparisons in relation to the whole, use a pie chart. If you want to add coherence to the narrative, the pie chart also helps because it depicts a whole. If your main interest is in stressing the relationship of one factor to another, use bar charts. If you wish to achieve all these effects, you can use either type of chart, and decide on the basis of which one is more aesthetically or pictorially interesting.

It is generally agreed that, in the United States, the eye is trained to perceive linear differences, such as those represented by bar charts, more accurately than differences represented by the arcs or angles of the circle. Nevertheless, there are many occasions when the pie chart is preferred; and accuracy can be preserved by showing exact amounts on the pie chart.

HOW MANY PIECES OF PIE?

Because the pie chart is used to dramatize relationships among segments and to the whole, it is important to keep the number of different segments to a minimum. Otherwise the reader will not comprehend the comparisons, and the desired effect will be lost. For the pie chart to be most understandable, it should in general have no more than eight segments. If the raw data you are using as bases for the pie chart have more than eight different factors, then group some together in order to keep the total at eight or fewer. Figure 6-3 shows three pie charts, one with four segments, one with eight, and one with twelve. You can judge for yourself the relative effectiveness of the pie chart as the number of segments increases. As with an apple pie, the more pieces you cut the less you get.

Figure 6-3

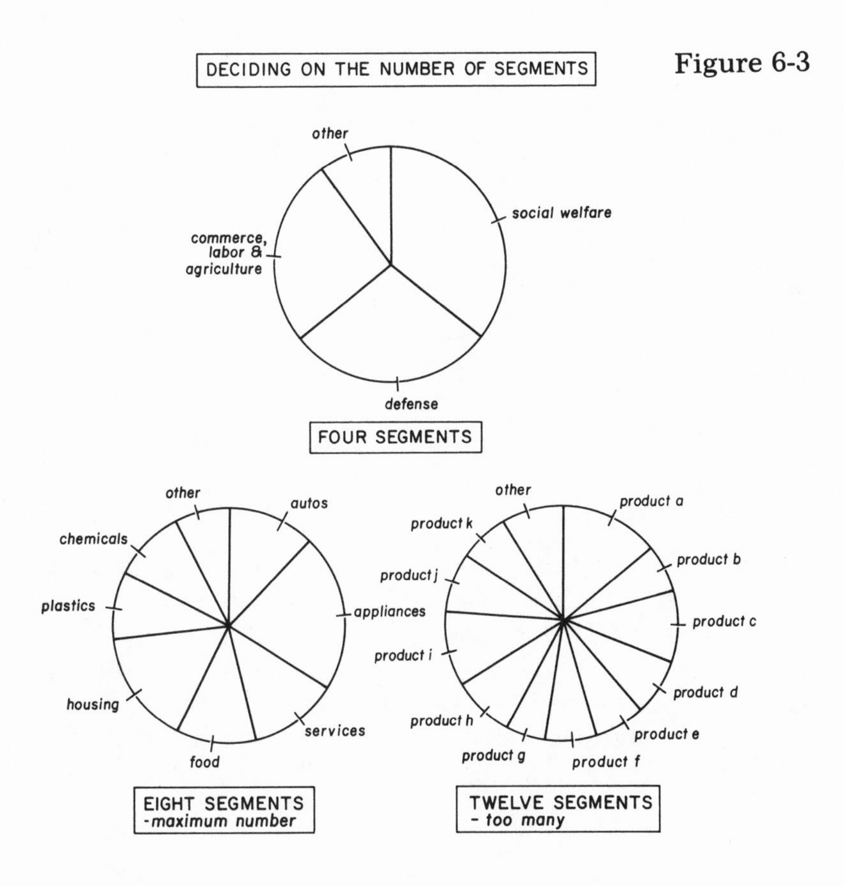

HOW TO PREPARE A PIE CHART

Many report writers shy away from using pie charts because of the difficulty in drawing circles and dividing them accurately. The following step-by-step guides provide a method to overcome this problem. Figure 6-4 is a working drawing of a circle, and it identifies the various components of a circle necessary to prepare the pie chart. It should be referred to in order to follow these instructions.

WORKING DRAWING OF A CIRCLE FOR USE IN PREPARING PIE CHARTS — Figure 6-4

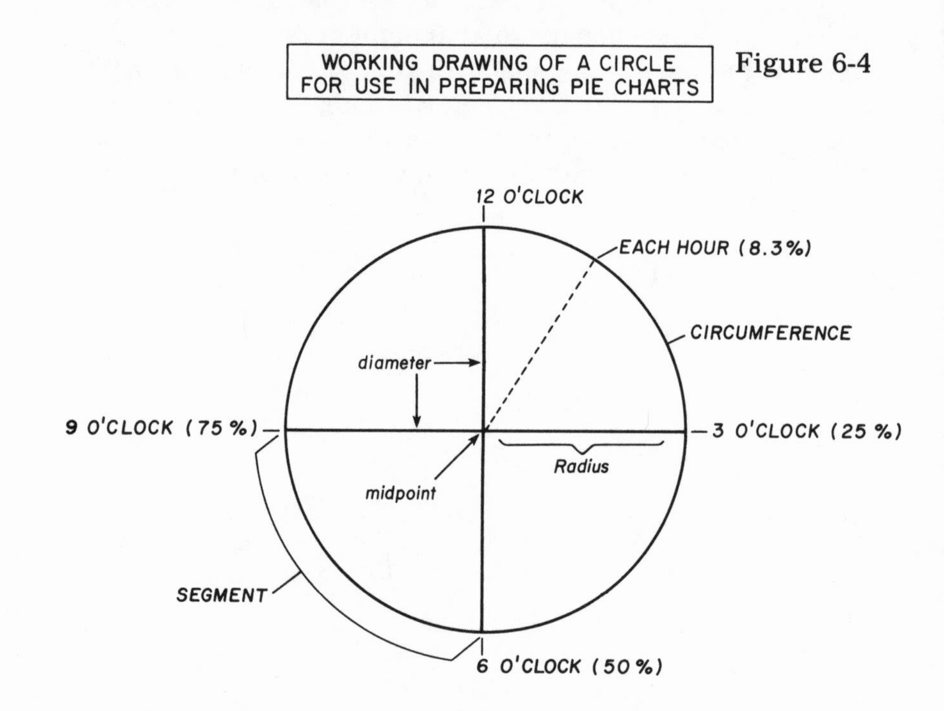

STEP ONE: Decide on the Size of the Circle

The initial step is to draw the circle, and the first question is: How large should the circle be? On standard 8½″ × 11″ paper the circle should be no less than 2″ in diameter. If it is smaller than 2″, there will not be sufficient room for the labeling. The chart will also lose its visual impact. If you can use a whole page or most of a page just for the chart, a better size is 3″ to 5″ in diameter.

Step Two: Draw the Circle

The best and easiest way to draw the circle is to use a plastic template such as Giant Circles (other brands are listed in Chapter 12, "Resources"). This template includes 11 circles of different sizes ranging from 1½″–3½″ in diameter. You place the template over your paper and trace the circle you want to use onto your paper. Most of these templates are also marked so that you can easily locate the centerpoint and the diameter of the circle. The diameter is the measurement across the center of the circle from one side to the other. It is the widest point in the circle. The line from the midpoint to just one side of the circle is called the radius.

Another method is to use a compass. A compass is helpful since it also marks the circle's centerpoint, which must be located in order to divide the circle into segments. In addition, compasses enable you to draw a circle of any size, whereas templates limit you to the sizes on the template. If you don't have a template or a compass, you can also draw the circle using a perfectly round glass or bottle of the desired diameter.

After you select the size circle you want, draw it on a piece of lined graph or cross-section paper. A pad of this paper can be obtained in art and office-supply stores. This is your worksheet to plan and draft the chart. Later you will redo or trace your illustration onto your regular report paper.

Think of the circle as the face of a clock with 12 o'clock at the top. It will make your work easier if you make sure that the top (12 o'clock) and one side (3 o'clock) both touch on cross-section lines of the graph paper.

Step Three: Locate the Diameters

With a pencil lightly draw in the horizontal diameter. This is a line that goes through the *center* of the circle from the left side to the right side of the circle—that is, it goes from 9 o'clock to 3 o'clock. Be sure the line goes through the center of the circle. You can locate the center by counting the number of squares on the graph paper from 12 o'clock to 6 o'clock and dividing in half.

Next, lightly draw in the vertical-diameter line from 12 o'clock to 6 o'clock.

STEP FOUR: Locate the Quadrants and Units of Measurement

Your circle is now divided into four quarters or quadrants. Each of these segments represents 25% of the whole circle. If you are thinking of your circle as a clock, every 3-hour segment is equivalent to 25%. Every 1-hour segment is equivalent to a measurement of 8.33%. Every 15 minutes is just over 2%. Still using a pencil, lightly mark off each of the 12 hours. Your diameter lines tell you exactly where 12, 3, 6, and 9 o'clock are. Using these as guides, you can estimate the location of the remaining hours.

STEP FIVE: Divide the Pie

The Clock Method / When dividing the circle it is usually most effective to start with the largest segment, beginning at

Figure 6-5

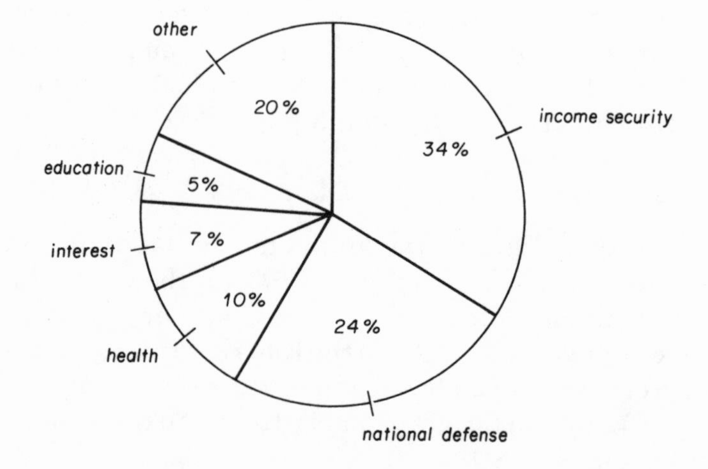

12 o'clock, and work clockwise (to the right) from the largest to the smallest segment. The exception to this order is if you have an "other" category. If you do, make the "other" segment the last one even though it is larger than some other segments.

Take the items of data you wish to depict and list them in descending order. The items used for Figure 6-5 are:

FEDERAL BUDGET, 1977	
Income Security	34%
National Defense	24%
Health	10%
Interest	7%
Education	5%
Other	20%
Total	100%

You are now ready to divide the pie. This is done by first drawing a line from 12 o'clock to the centerpoint. The next line is drawn from the centerpoint to the point on the circumference (that is, the approximate hour) that represents the value of the segment. The first value in this example is 34%. You know 25% is at three o'clock. From 3 o'clock to 4 o'clock is another 8.33%. Thus your line should be drawn to about 4 o'clock (25% + 8.33% = 33.33%).

The next segment is to represent 24%. You need to add a little less than 3 additional hours. (3 × 8.33 = 24.99%). Adding the 3 hours to 4 o'clock brings you to about 7 o'clock. Draw the line from the midpoint to 7 o'clock. Continue on until all segments are shown. Adjust the lines a little if the chart doesn't seem to be coming out right. Remember: You do not need to be exact. Use a ruler or edge of the template to be sure your segment lines are straight.

The Protractor Method / If you want to be more exact in your measurements, you can use a protractor. A protractor is a metal or plastic half circle that has the circumference marked off in degrees. There are 180 degrees for a half circle, 360 degrees for the whole circle; 1% is equivalent to 3.6 degrees. To mark off 34%, for example, you would measure 34 × 3.6 or 122.4 degrees.

STEP SIX: Transfer the Chart

After you complete the chart on your worksheet, draw it again on the regular paper using the worksheet as a guide. When you have prepared a few pie charts using the above method, you may find that you don't need to use graph paper as a worksheet but can use plain paper and then redo the draft on regular report paper.

LOCATING ABSOLUTE NUMBERS

The preceding examples are based on using percents. Suppose, however, that you want to show absolute whole numbers in a pie chart. How do you figure out the size of the segments? This is done by converting the numbers to percents and then locating the segments, as above. For example, a set of information on ethnic characteristics of a population is as follows:

White	4,000
Black	2,000
Hispanic	1,500
Other	1,000
Total	8,500

To figure the percents divide the total, 8,500, into each number and you get:

White	47.1%
Black	23.5%
Hispanic	17.6%
Other	11.8%

Use these percents as the bases for dividing your circle into segments. If you wish, you can label each segment of the chart using the absolute number even though you used percents to locate the segments. This labeling option is illustrated in Figure 6-6.

LABELING WITH WHOLE NUMBERS

Figure 6-6

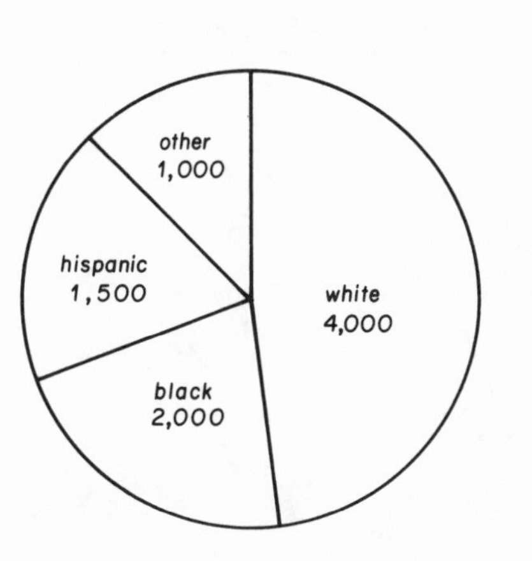

SHADING PIE CHARTS

The pie chart is quite effective without shading. As long as it is properly labeled the comparisons show up quite well. However, shading adds additional interest and also further differentiates each segment of the circle. An example of a shaded pie chart is shown in Figure 6-7.

Shading for a pie chart is done in a manner similar to the bar chart. The best method is to use self-adhering shading sheets. Cut out the correct size to cover each segment, remove the backing, and press the piece on. Another method is to rub a pressure-sensitive shading sheet with a stylus or pencil; the shading comes off on the chart. A listing and description of this material and further instructions on its use appear in Chapter 12, "Resources." Shading can also be done by hand. Avoid doing it with a typewriter or charting tapes—these work satisfactorily with the straight lines of a bar chart, but are difficult to use with the angles of a pie chart.

All the segments do not have to be shaded, nor is it necessary to use a different shading pattern for each section (see Figure 6-7). Because each shaded segment is almost always labeled, it is usually not necessary to accompany the shaded pie chart with a key or legend. On some occasions when the labeling is too long to show on the chart you may wish to use a key.

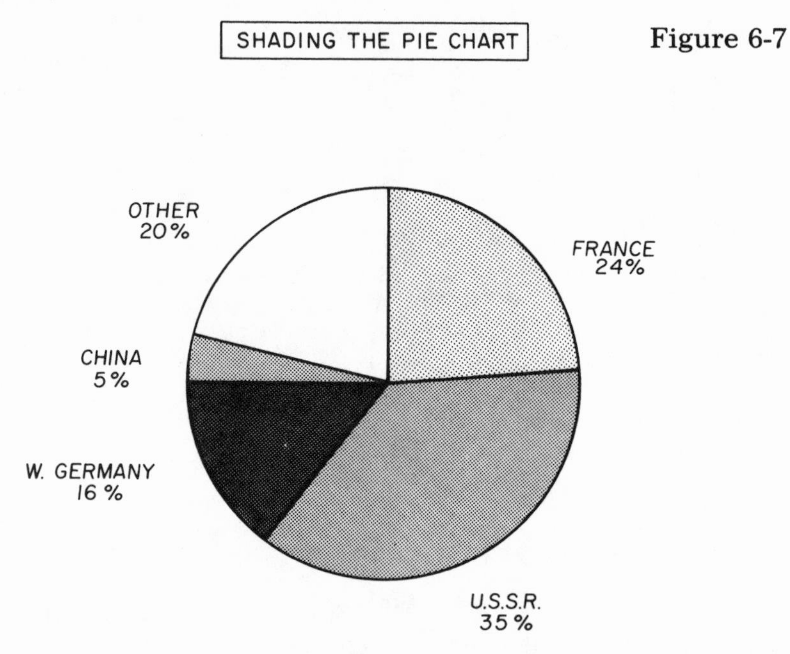

Figure 6-7

LABELING THE PIE CHART

Labeling for the pie chart should include:

- a title above the chart that includes the chart number; the factor being analyzed (for example, imports, federal expenditures, sex, age, race, etc.); the year or years for which the chart pertains; and, where necessary, the unit of measurement (for example, number, percent).
- a label for each segment to show the *specific* descriptive unit of analysis that refers to each segment. For example, if the unit of analysis in the title is race, then each segment would be labeled white, black, Hispanic, etc.
- an additional option for each segment's label is to show the percent or number that the segment represents, such as "women (35%)" or "black (2,500)."

Descriptive labeling for the pie chart can be done within the appropriate segment or adjacent to the segment. Often there is not enough room to label within the segment, and the label must be outside the circle. Be sure it is next to the correct segment. If there is not room to get it right next to the segment, use a line or arrow to connect it to the proper segment of the chart.

Another alternative is to put the specific unit of analysis label adjacent to the segment and put the appropriate measurement (that is, the number or percent) within the segment. Numbers take up much less space than descriptive words. If you have shaded the segments, it will be much easier to place all labels outside the segments.

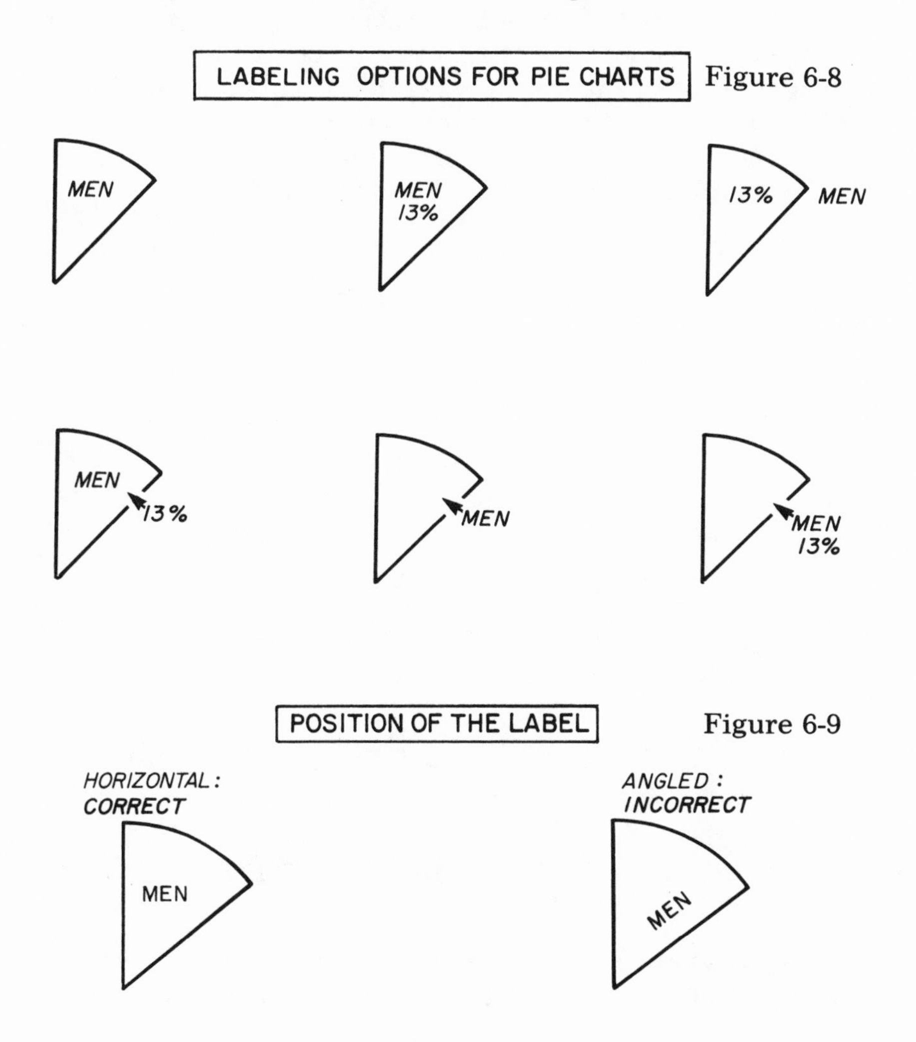

LABELING OPTIONS FOR PIE CHARTS Figure 6-8

POSITION OF THE LABEL Figure 6-9

These various alternatives are shown in Figure 6-8. Make sure that all lettering, whether done by hand or on the typewriter, is perfectly horizontal. It is a common mistake to have the lettering within a segment follow one of the angled lines that divides up the circle. Figure 6-9 shows improper and proper positioning for labeling the pie chart.

PIES AS COINS

One of the most common uses of a pie chart is to depict financial information, usually by showing how a total amount of money is divided up. When pie charts are used to represent expenditures or income, the chart is often drawn as a coin, as in Figure 6-10. The disc-like effect is accomplished by moving the circle template down a fraction of an inch and drawing in the second circumference line along the bottom and then adding the vertical ridge lines.

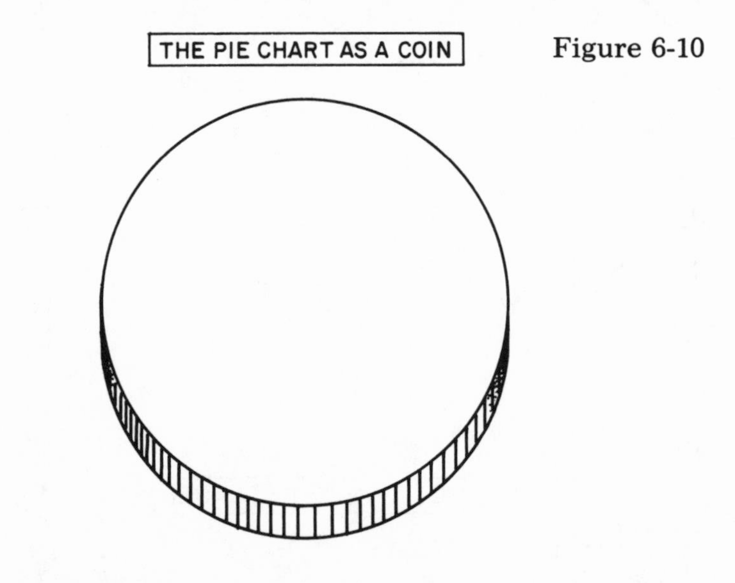

Figure 6-10

INCREASING THE COMPARATIVE POWER OF THE PIE CHART

Pie charts can be used to show comparisons by placing two or more pie charts next to each other, as in Figure 6-11.

Notice that the charts depict the same type of information, the circles are all the same size, and they all represent 100%. But since the charts show measurements for different years, the segments are of different proportions.

COMPARATIVE PIE CHARTS

Figure 6-11

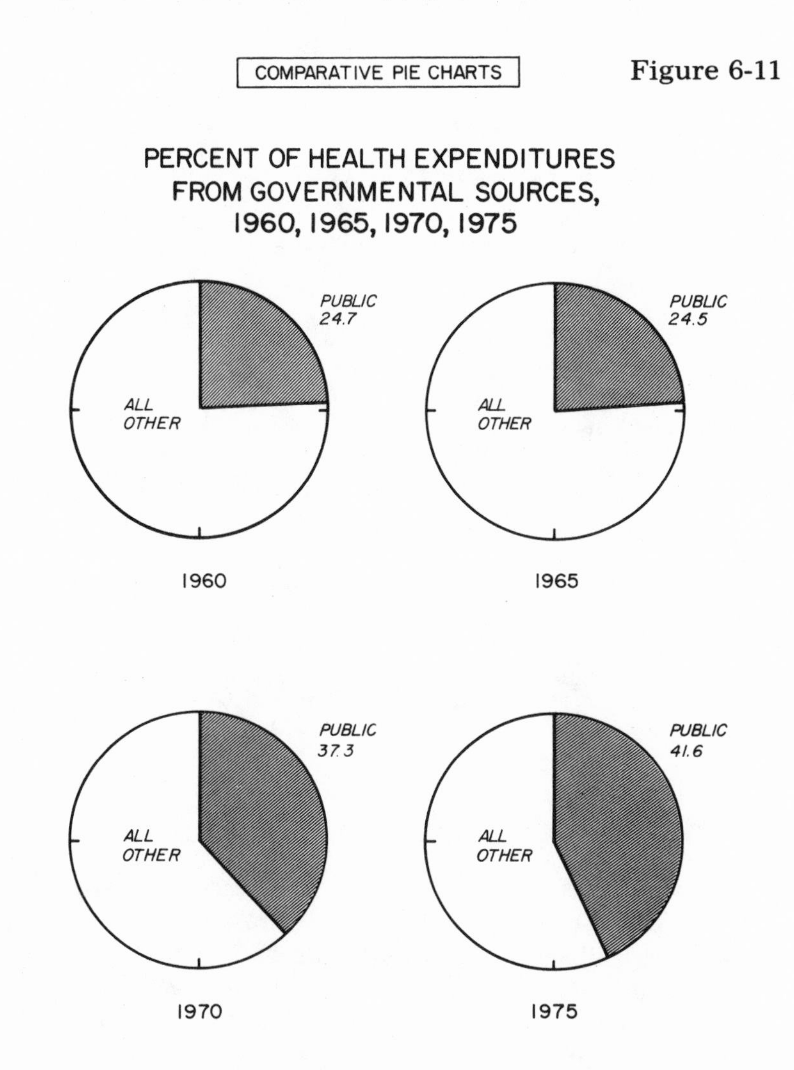

It should be noted that comparisons with a circle generally appear to show less difference than bar-chart comparisons. The larger the circle you use, however, the more these comparisons will show up. And shading of segments helps to emphasize differences.

USING SEPARATED SEGMENTS

Figure 6-12

FOREIGN POLICY VIEWS

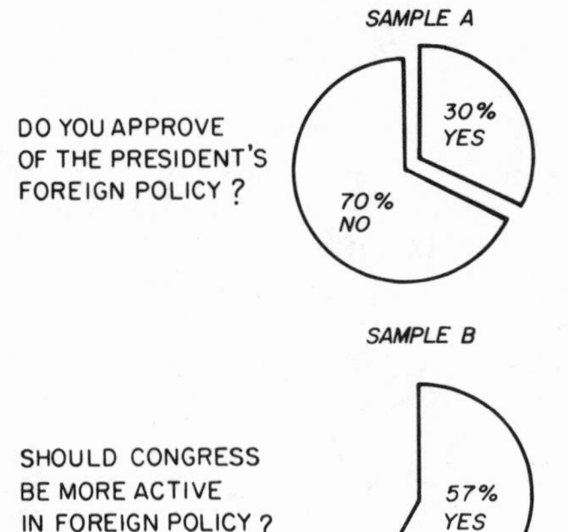

THE USE OF SEPARATED SEGMENTS

One of the variations of the basic pie chart is to emphasize an individual segment of the chart. Emphasis is achieved by separating that segment from the total chart, as in Sample A of Figure 6-12. This is harder to draw than the basic pie. It can be accomplished in two ways. One is first to draw the total pie and its segments. Then trace the pie without the emphasized segment. Trace that segment separately. Now cut out the two tracings and cement them to a piece of plain white paper with the one segment placed in its separated position. When you photocopy this figure, it will look as if it was originally drawn this way.

The other method, which is much faster and more expedient, is first to draw the segment using the circle template and then move the template down and a little to the left and draw in the rest of the circle. This will give the appearance of Sample A in Figure 6-12.

A further adaptation of this variation is to use only the segment, as in Sample B of Figure 6-12. You can also shade the segment for additional interest and emphasis, as in Sample C of Figure 6-12.

ADDING INTEREST TO THE PIE CHART

Using a background such as the lines of graph paper or placing the pie in a box are other techniques to add interest to the chart and contribute to the overall impact of the report. Boxing the circle also helps to anchor the circle to the page. In

ADDING A GRID FOR INTEREST   Figure 6-13

STATE BUDGET INCOME SOURCES,
F.Y. 1980-81

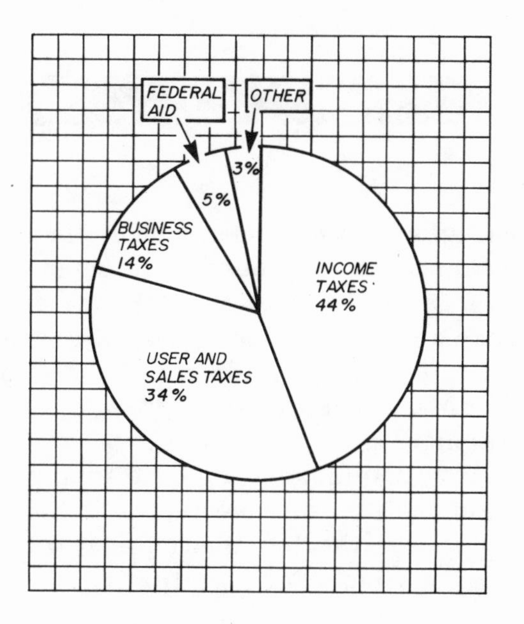

addition, you can use shadings to give a three-dimensional appearance to the chart. Some examples of these techniques are illustrated in Figures 6-13 and 6-14.

ADDING BOXING AND DIMENSION

Figure 6-14

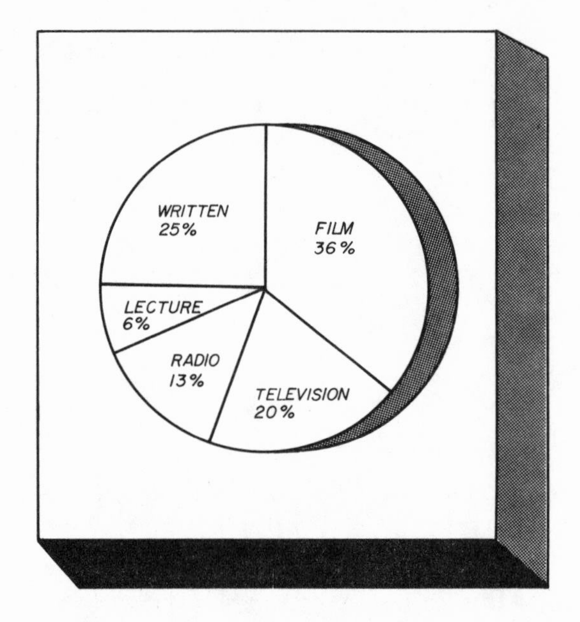

Another variation is to place a second circle in the middle of the pie to show the total amount represented by the pie, as in Figure 6-15. Draw the total outside circle and mark the midpoint. Take a smaller circle and draw the inner circle, being careful to match the midpoint of the smaller circle with the larger one. Then proceed to draw the segments and label. Another method is to draw the regular chart with all its segments and then make a separate smaller circle that you cut out and paste on in the center of the larger circle.

The pie chart and the bar chart are two of the basic graphic forms for comparing quantitative information. In

Figure 6-15

SHOWING TOTAL AMOUNT AND PERCENTS

STATE EXPENDITURES 1980-81

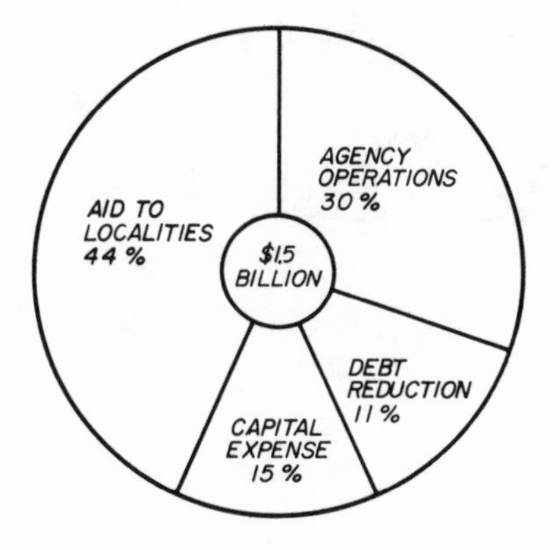

the next chapter we will explore a third popular method for depicting quantitative information: the line chart or line graph.

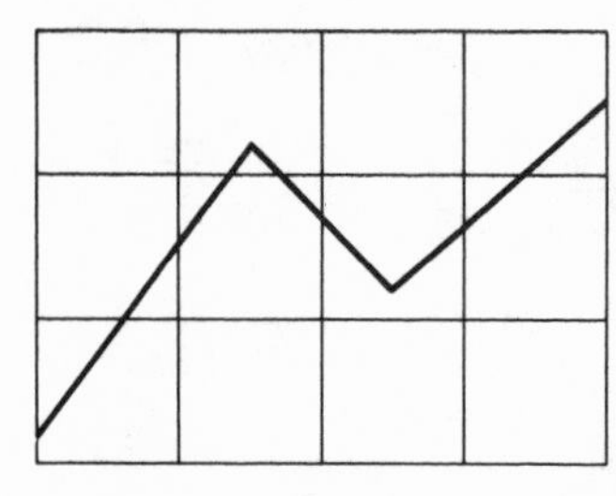

CHAPTER SEVEN

The Line Chart

WHAT IS A LINE CHART?

The line chart is a third type of graphic that can be used to describe and compare numerical information. The line chart is also known as a line graph, and it has a variety of other names, such as "curve chart" and "curve graph." Like the bar chart and the pie chart, the line chart enables you to portray quantitative information in an interesting and efficient manner. There are many adaptations of the basic line chart that have specific purposes depending on the types of scales that are used or the way in which the shape of the plotted line is drawn. These include frequency-distribution charts or frequency polygons, step charts, normal curves, ogives, bell-shaped curves, scattergrams, surface graphs, and others.

The two most frequently used types of line charts in the kinds of reports with which this book is concerned are (1) time charts and (2) frequency distributions. Most of the discussion in this chapter is focused on these two major forms.

The basic line chart is comprised of a horizontal axis, a vertical axis, ruled grid lines to show the measurement scales, and a plotted line or lines. Each axis contains a measurement scale that shows age groups, amounts, percents, rates, scores, income groups, time periods, and the like. Each plotted line represents the value of some factor that is being measured. For example, changes in the population over a period of time may be shown by using a line graph in which: (a) time periods, expressed by a series of years, are used as the scale of measurement along the horizontal axis, and (b) numbers of people (another scale of measurement) are used along the vertical axis. The plotted line represents the relationship between these two measurements—that is, it shows the relationship of the number of people in the population for each of the designated time periods, as in Figure 7-1.

A BASIC LINE CHART

Figure 7-1

POPULATION OF MIDVILLE : 1945-1980

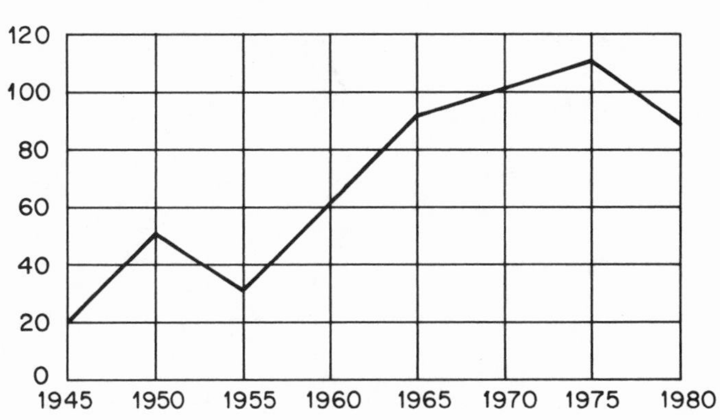

WHAT DO LINE CHARTS ACHIEVE?

Line charts are most useful to depict three different types of information:

- *trends* or *changes* over time—for example, change in population, as in Figure 7-1, or in an organization's income over a twenty-year period, as in Figure 7-2.
- the *relationship* between the distribution or occurrence of two quantities or variables—for example, the distribution of the number of participants in a program according to their ages, as in the frequency-distribution chart shown in Figure 7-3.
- the *comparison* of both trends and relationships—for example, Figure 7-4, which enables us to compare trends in per-capita income for three different cities over a period of forty years.

SHOWING TRENDS

Figure 7-2

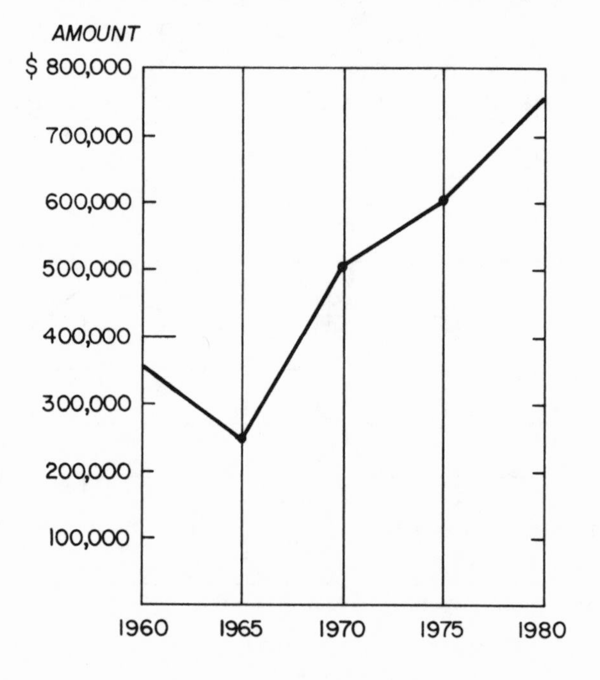

The line chart is one of the oldest and most frequently used graphics. It has a number of quite technical applications in the field of statistical analysis. In Chapter 12, "Resources,"

Figure 7-3

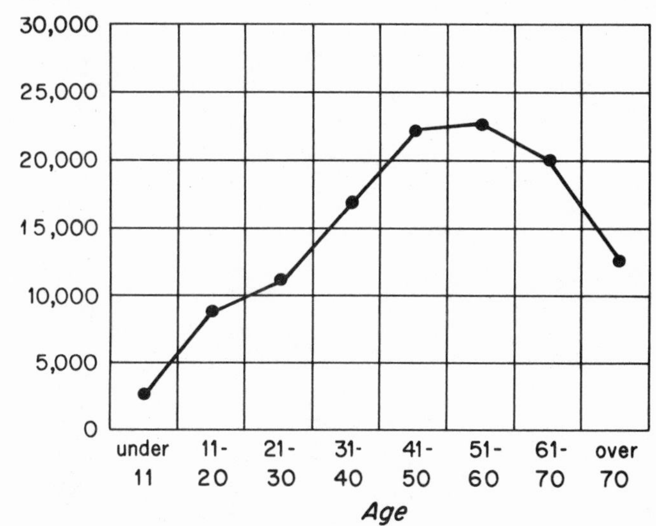

there is a list of resource books that can be used to pursue further the technical aspects of the line chart if you wish. Our concern is with the line chart as a tool for explanation and persuasion. In this regard the line chart can be used to achieve the following objectives in a report:

- ▶ summarization
- ▶ coherence
- ▶ interest
- ▶ impact
- ▶ credibility

The line chart is a very efficient graphic. It can be employed to present considerable quantitative information in a form that enables the reader to see trends and relationships quickly. Because it can be used to summarize information

COMPARING TRENDS AND RELATIONSHIPS

Figure 7-4

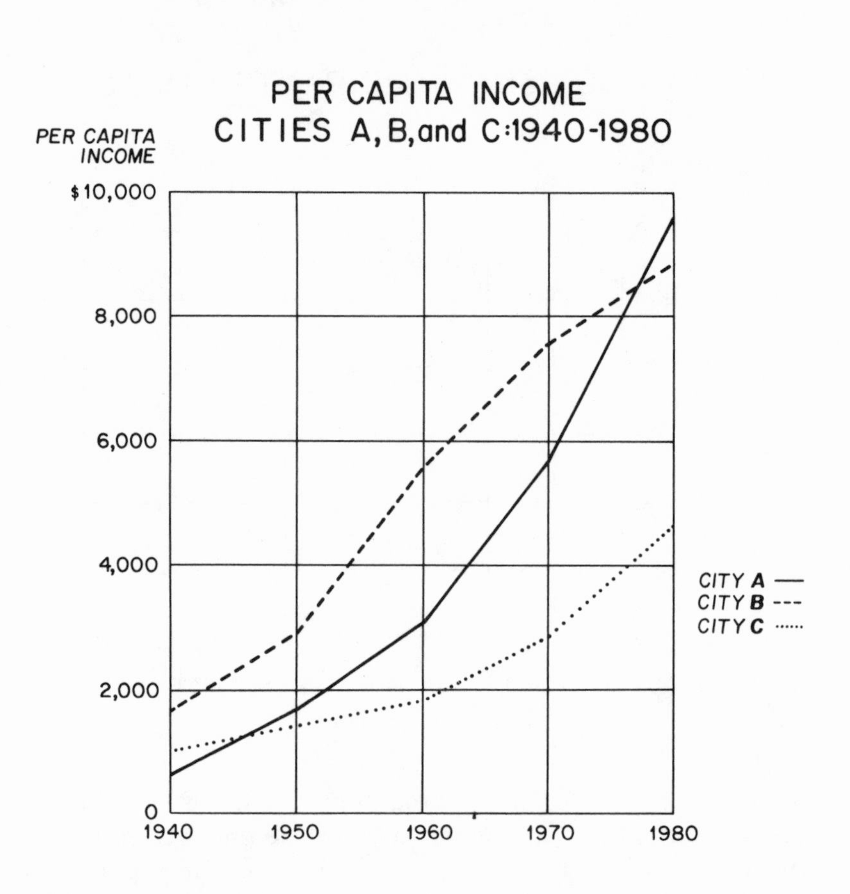

from rather extensive tables and present it in a graphically understandable way, it adds interest and impact to a report. Further, use of the line chart also lends a degree of credibility to your presentation, since readers often associate it with science and technology. While it is used in some reports as a substitute for a table, it is frequently seen as a supplement to the table. If you don't want to put the table in the text, you can include it in an appendix and make reference to its location in the text.

Line charts should be used sparingly. Most readers resist page after page of any kind of chart or graph, but especially line charts. This is true because the line chart frequently (although not necessarily) shows much more information than the bar chart and the pie chart, and you must be careful to avoid placing an informational overload on the reader.

STRAIGHT LINES OR CURVES?

Everyone has seen line charts that either use jagged lines, straight lines, curves, or steps, as illustrated in Figure 7-5.

Figure 7-5

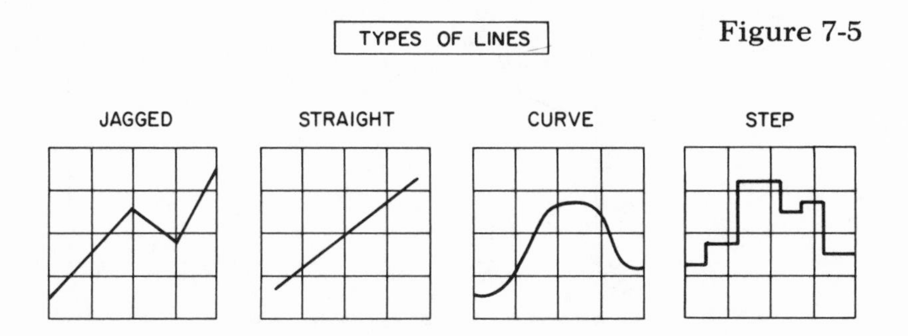

Which one to use? There are certain rules in technical statistical presentation that apply to the use of straight lines or curves, but they have limited relevance for our purposes. (See Chapter 12, "Resources.")

More pertinent considerations related to report preparation have to do with which form is more effective and maintains the integrity of your presentation. Most of the data included in business, professional, governmental, and organizational reports lend themselves to the use of the jagged lines. These are also referred to as zigzag lines. They are much easier to draw and to read than curves. Thus, with a few exceptions, the illustrations in this chapter will not make use of curves.

COMPONENTS OF THE LINE CHART

Figure 7-6 shows the components of the line chart. You will notice that it has some similarities to the basic form used for the bar chart described in Chapter 5. The five principal line-chart components are (1) a vertical axis, (2) a horizontal axis,

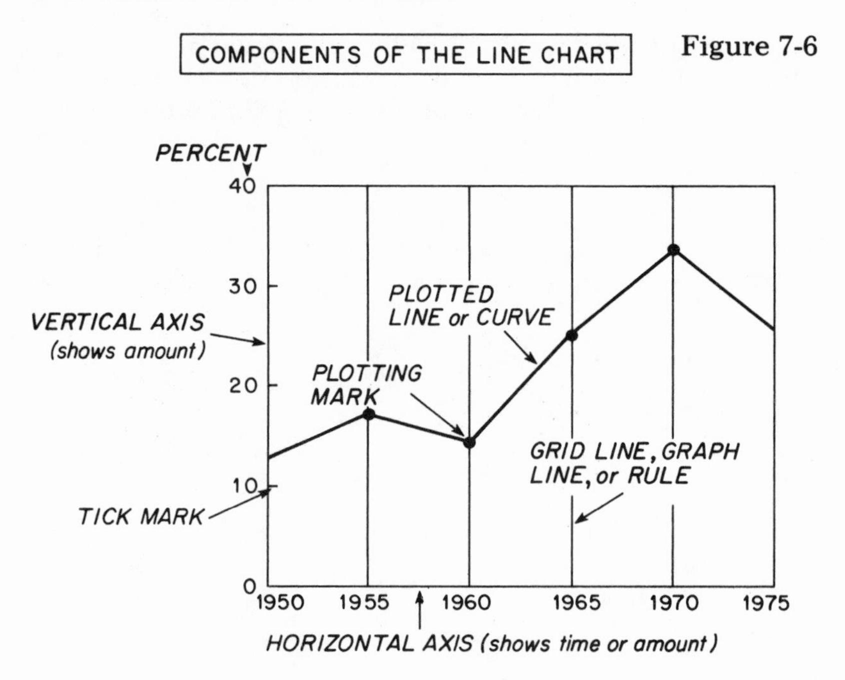

Figure 7-6

(3) the plotted line or curve, (4) plotting dots or marks (optional), and (5) scale or grid lines. (The vertical axis is also called the y axis or ordinate, and the horizontal axis is also called the x axis or abscissa.)

Unlike the bar chart, the line chart makes use of two scales of measurement. The line chart has one scale of measurement along the horizontal axis and another scale of measurement along the vertical axis. This means that each plotted line shows the relationship between whatever these two measurements represent. For example, in Figure 7-6 the horizontal scale is expressed in years as one unit of measurement. The vertical scale measures the amount, in this case expressed in percents. The plotted line shows the amount (percent) for each time period (years).

In presenting statistical material, the measurement of the independent variable is usually placed on the horizontal axis, and the dependent variable on the vertical axis. Another way to think of this construction is to view the vertical axis as measuring frequency (or amounts) and the horizontal axis as designating the method of classification.

SHOWING UNITS OF MEASUREMENT

In order to help the reader better understand the actual values that the plotted line represents, it is necessary to include either tick marks or the more extended grid lines (also called graph lines or rules) for both scales. As shown in Figure 7-7, it is possible to use all tick marks, all grid lines, or combinations of the two.

Figure 7-7

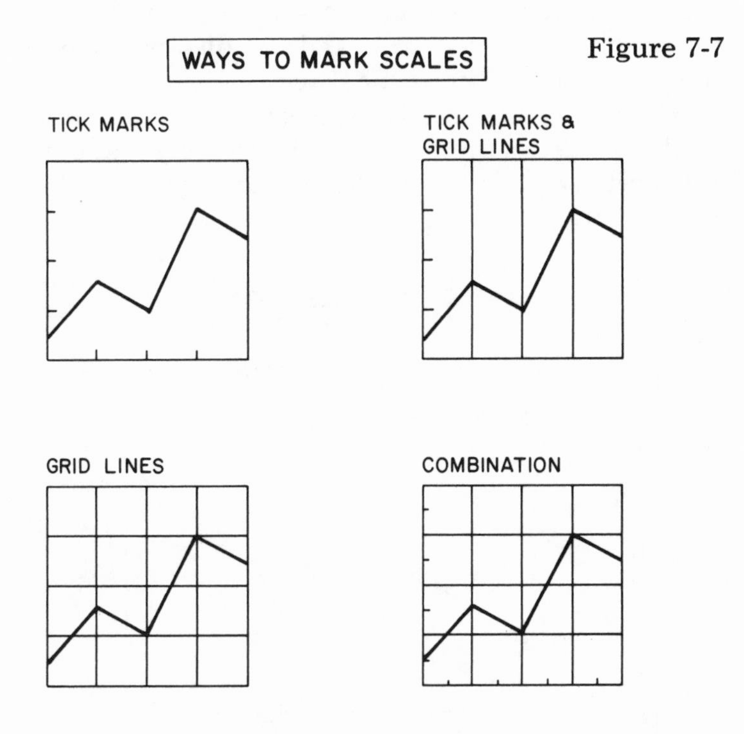

The units of measurement that are used on the horizontal scale and on the vertical scale must all be the same distance apart on each scale. (An exception is in the case of what are known as proportional-line charts, but these require quite complicated scaling and are not included in this discussion.) While each scale must be internally uniform, the two scales do not have to match—that is, the units on the vertical scale may all be ¾″ apart, and the units on the horizontal scale may all be ½″ apart.

SCALE PROPORTIONS

Selecting the distances for the units of measurement is crucial in presenting line graphs since it is possible to change the effect of the graph completely by selecting different scale amounts and distances. The two graphs in Figure 7-8 both present the same information but use different scales. The result is two very different impressions. The first shows considerable change from year to year; the second minimizes the impression of change. This is the result of altering the length of the vertical axis and the distance between each measurement on the vertical scale.

Which more accurately depicts the "real" situation? In the field of statistics there has developed a general guideline or convention known as the "three-quarter rule." This rule sug-

Figure 7-8

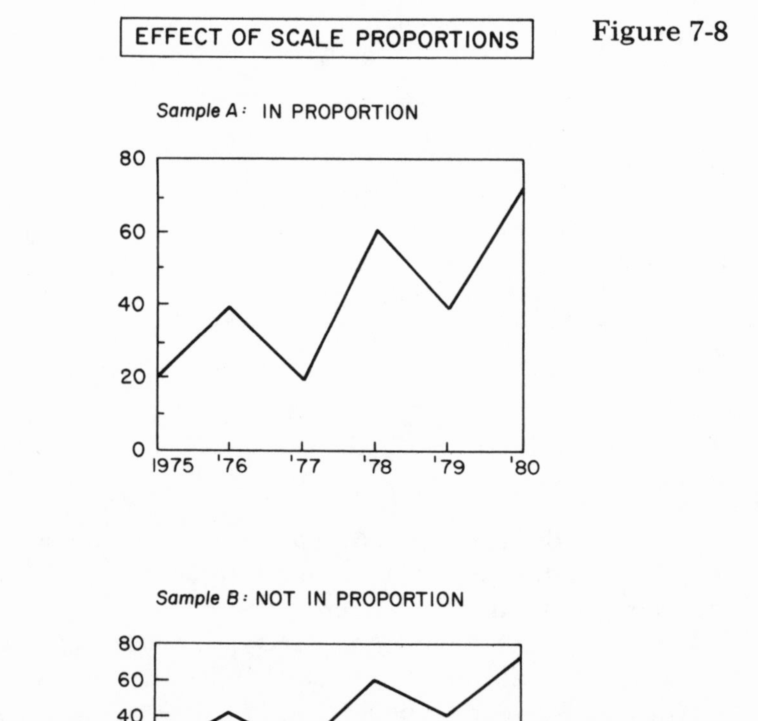

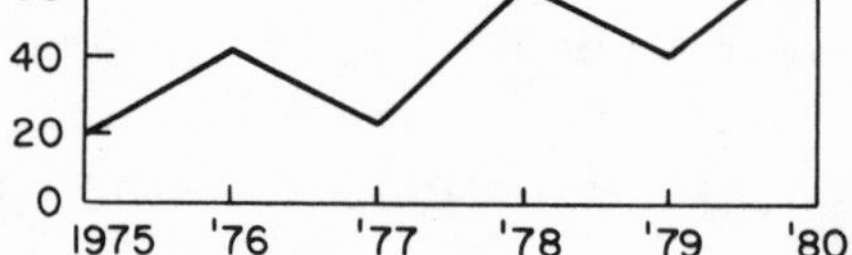

gests that the height of the vertical axis should be about three quarters the length of the horizontal axis. If the chart does not lend itself to this scale, another general guideline is to keep the chart in approximate proportion to the page dimensions. As a further guide to eliminating distortion, all amount scales should start from zero. Sample A in Figure 7-8 follows these guidelines; Sample B does not.

To reduce the distance of the amount (vertical) scale tends to minimize differences; to expand this scale by making it longer and making the amount markings farther apart will maximize differences. One has considerably more latitude in devising scales that will emphasize or de-emphasize information when using line charts than in the case of pie charts. Scales for bar charts can also be manipulated, and in order to preserve the integrity of the presentation, the same general guidelines should apply to them as to line charts.

HOW MANY PLOTTED LINES?

Line charts can have just one plotted line or may have several lines. Each additional line increases the information shown in the chart and enables the reader to see both *relationships* and *comparisons.* In Figure 7-4, referred to earlier, the trend in per-capita income for three different cities is depicted. Thus one can observe the *relationship* between income and years for each city and can also *compare* the three cities to each other in this regard.

For most line charts the maximum number of plotted lines should not exceed five; three or fewer is the ideal number. When multiple plotted lines are shown each line should be differentiated by using (a) a different type of line and/or (b) different plotting marks, if shown, and (c) clearly differentiated labeling. Later in this chapter examples of the variety of different lines and marks that can be used are presented.

TIME CHARTS

Line charts that show changes or trends over time are called time charts. A time chart is one of the most frequent applications of the line chart. When you use the line chart for this

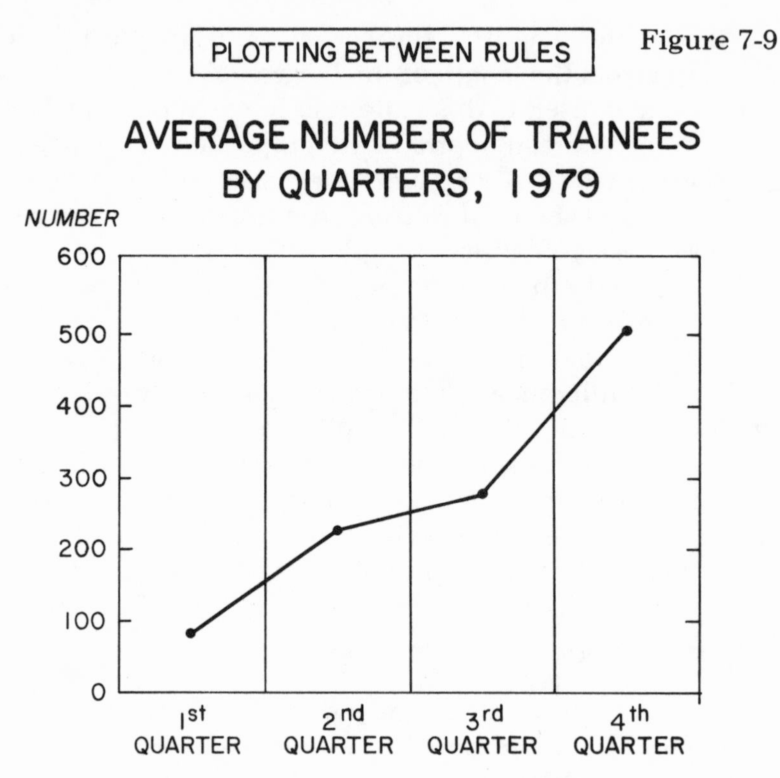

Figure 7-9

purpose the vertical axis should be the amount scale. These amounts can be expressed in percents, absolute numbers, or rates. The scale should begin with a base of zero and go up in equal steps or units. The horizontal axis should show the scale of time periods. Time periods should be expressed in equal units such as one year, five years, or ten years. The distance between each unit on the scale must be equal.

Once you have your amount scale and time-period scale devised, the next step is to plot the line by putting in the plotting dots or other marks for each time period. The location of these dots in relation to the amount scale is easy to determine. The amount scale is really a vertical linear continuum of amounts. The plotting dots are located along this continuum. Location of the dot in relation to the time scale, however, requires additional planning. This is because the amount measured is either for an entire period (for example, 1965) or for a given date (for example, December 15). Where the amount to be shown is for an entire time period such as a month, a quarter, a year, or a five-year period, the plotting dot may be placed *on* the vertical time-scale lines as shown earlier in Figure 7-6; or it may be centered *between* the lines, as in Figure 7-9. Whenever possible it is preferable to plot on the lines since this is both easier to prepare and to read accurately.

Because of space limitations it is often necessary to use abbreviations when designating each time unit of the horizontal scale. Years, for example, can read '69, '70, etc. Months can be Jan., Feb., etc., or you can just use letters J, F, M, A, etc.

FREQUENCY-DISTRIBUTION CHARTS

The frequency-distribution chart is very similar to the time chart except that the horizontal scale is not based on a series of time periods. Instead it shows another measurement factor (or variable) expressed as a quantity being measured, units produced, etc. Whatever the characteristic, it must be expressed in units of the same interval. For example, if you are going to use age groups, they should all be for the same number of years such as 0–4, 5–9, 10–14, 15–19, etc., all of which represent 5-year intervals. Or, if you are showing number of units produced, the horizontal scale might be 0, 100, 200, 300, 400, 500, etc.

The vertical axis of the frequency-distribution chart is scaled to show the frequency of occurrence by using an amount scale as in the time chart explained earlier. This can be expressed in numbers of persons, in dollars, in percents, or in any other numerical unit that can represent the frequency of occurrence. The plotted line then would show the frequency with which each of the units or characteristics on the horizontal axis occurs. An example of a frequency-distribution chart that shows how frequently persons in different age groups participated in a program was shown earlier in Figure 7-3. One of the typical uses of the frequency-distribution chart is to show the distribution of test scores, as in Figure 7-10 for all groups and for different groups (in this case males and females).

Figure 7-10

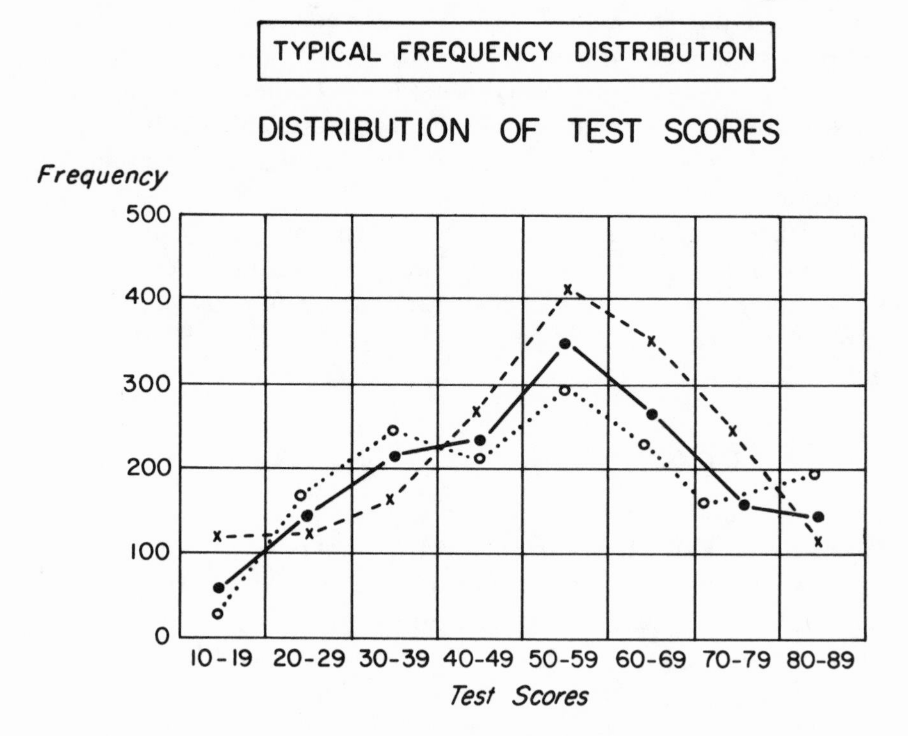

HOW TO PREPARE A LINE CHART

There are five major steps in preparing a line chart, as follows:

STEP ONE: Decide on the Size of the Chart

Line charts occupy a square or rectangular area of the page, and the first step in preparing the chart is to decide on the amount of space you want to devote to the chart. Assuming the use of 8½″×11″ paper and typewritten reports rather than printed ones, the following table and Figure 7-11 provide you with guidelines for making this decision:

	HORIZONTAL AXIS		VERTICAL AXIS
Maximum	7½″	×	8″
Minimum	3″	×	2″

The maximum size chart (7½″×8″) should occupy an entire 8½″×11″ page in order to leave room for the title and labeling; there would not be room for any other copy. The minimum size would enable you to have about one-half page of narrative copy on the same page.

Figure 7-11

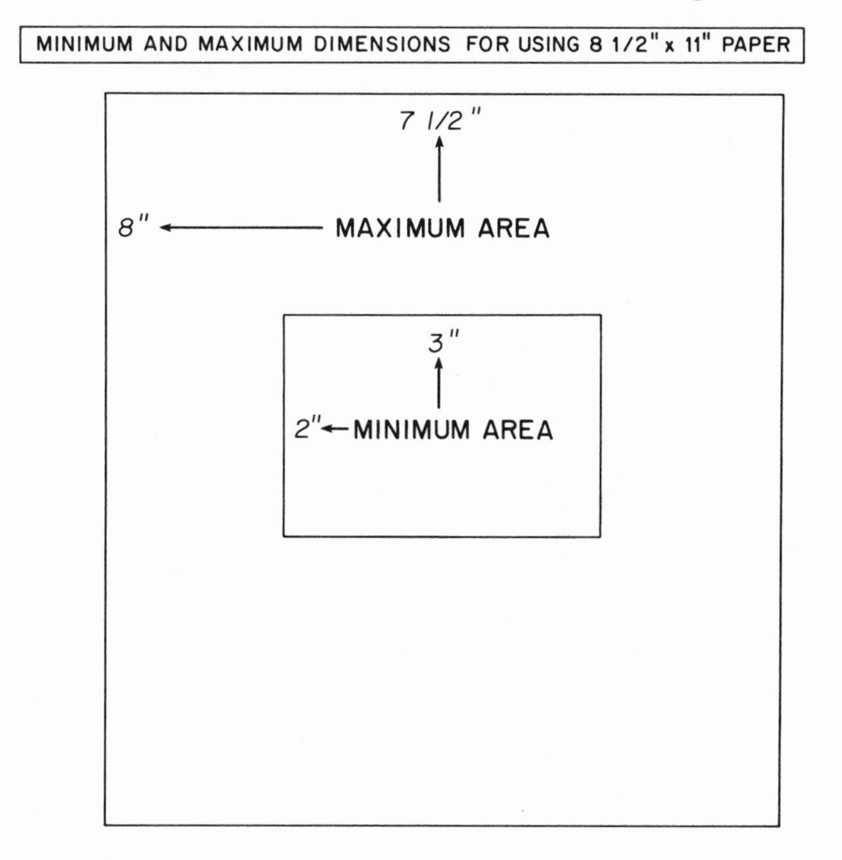

STEP TWO: Decide on the Units of Measurement and the Scale

The next step is to decide what units of measurement to use on the vertical axis and on the horizontal axis. Since most line charts are based on data that are included in a table or tabulation, the units of measurement are, of course, based on those used in the table. They may be the same as those in the table; or may be just certain ones selected from the table; or may represent some regrouping of those in the table. For example, Table 7-1 is a set of data that are to be shown in a line chart.

You can devise a vertical and a horizontal scale that correspond to each of the raw-data units shown in Table 7-1, and each item in the table is plotted. The result will be a chart such as Example A in Figure 7-12.

Figure 7-12

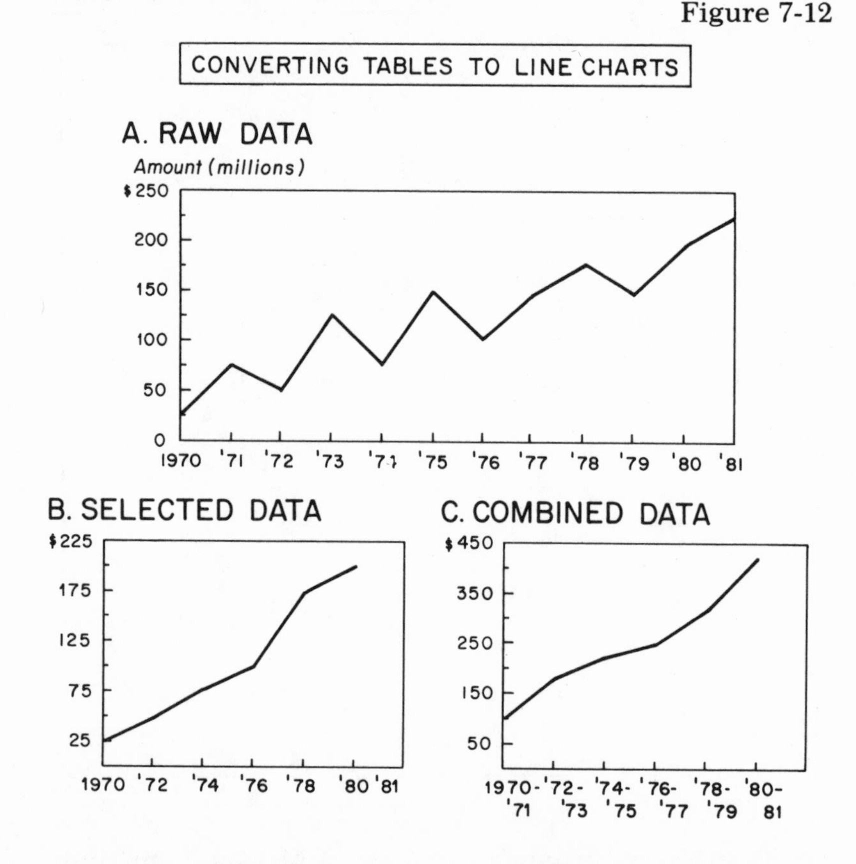

Alternatively, you may wish to simplify the information and *select* data from Table 7-1 that results in a new table, Table 7-2. This would require a change in the horizontal scales and a chart like B in Figure 7-12.

Or you may simplify the data by combining or classifying the data, as in Table 7-3. Combining data in this manner would result in a chart like C in Figure 7-12.

Table 7-1 ORGANIZATION EXPENDITURES FOR HEALTH PROGRAMS

YEAR	AMOUNT (IN MILLIONS)
1970	$25
1971	$75
1972	$50
1973	$125
1974	$75
1975	$150
1976	$100
1977	$150
1978	$175
1979	$150
1980	$200
1981	$225

Table 7-2

YEAR	AMOUNT (IN MILLIONS)
1970	$25
1972	$50
1974	$75
1976	$100
1978	$175
1980	$225

Table 7-3

YEAR	AMOUNT (IN MILLIONS)
1970–71	$100
1972–73	$175
1974–75	$225
1976–77	$250
1978–79	$325
1980–81	$475

Since many line charts, especially time charts, do not show the exact measurements, it is often desirable to combine and simplify information as much as possible. However, you must decide how general or nonspecific you want to depict the information and the extent to which simplification and selectivity may tend to distort the information. The charts in Figure 7-12 all are based on the same data. Example A emphasizes an up-and-down pattern. Example B shows a steady but not smooth increase. Example C shows a smoother increase.

STEP THREE: Lay Out the Chart Scales

The third step is to lay out the chart in draft form. The most effective way to do this is to use printed graph paper and then redo the chart on white typing paper.

Graph paper is ruled with both vertical and horizontal lines. It comes in a number of different rulings, but the easiest to use are those that have four to six rules (or boxes) per inch.

Prepare the Horizontal Scale / Start your layout by preparing the horizontal scale along the bottom of the chart. Do this by using the following procedure:

1. Decide on the number of units of measurement or classification you will have. For example, assume you are going to show organizational expenditures for each year over a twelve-year period based on the data in Table 7-1. This will require twelve equidistant units, each of which represents a year.
2. Decide how many of the ruled lines on the graph paper you want to use for each unit (year). In the original drawing for Example A in Figure 7-12, two ruled lines were used for each one-year period when this was first laid out on graph paper having four lines to the inch.
3. Draw in the horizontal-scale line, being sure to put it low enough on the page to leave room for the vertical scale and the chart number and title.

Prepare the Vertical Scale / Follow the same procedure to prepare the vertical-amount (or frequency) scale. Each step in this scale must also be equidistant but may be spaced differently than the horizontal scale. In Example A, Figure 7-12, the amounts are shown in steps of $25 million. One space of the graph paper was used in the original drawing for each step. In Example C, two spaces were used for each $50 million step. Always try to select intervals that will help give the chart a symmetrical appearance—that is, it should not have a severe oblong appearance, but should be more square and related to the dimensions of the 8½″×11″ page.

If you do not use graph paper, there are rulers used for drafting that you can buy in office-supply stores. These rulers have, in addition to inches, decimal scales that have ten to sixty marks per inch that you can use to mark off a uniform scale along the sides of your chart. In addition, the computer template referred to earlier has different scales marked along its edges, which you can use to measure off scales.

Draw in the scale lines using one of the alternatives shown earlier in Figure 7-7 (tick marks, full scales, or a combination of tick marks and full scales).

STEP FOUR: Prepare the Plotted Line

Having prepared the horizontal and vertical scales, you are now ready to draw in the plotted line or lines. This is done by entering plotting marks or points on the graph and then joining each mark with a straight line or one of the other types of plotting lines shown in Figure 7-13. The location of each mark is defined by the data set from which you are working. Plotting marks may be placed on the scale line or centered in the space between the scale lines. As pointed out earlier, it is much easier and clearer to plot on the scale lines. You may either show the plotting marks in your final drawing or eliminate them.

If you want, you can draw the plotted line as a curve. Figure 7-14 shows the same data as Example A in Figure 7-12, except that the line is drawn as a curve by rounding the line as it approaches and goes through the plotting points. (It should be noted that from a technical statistical point of

Figure 7-13

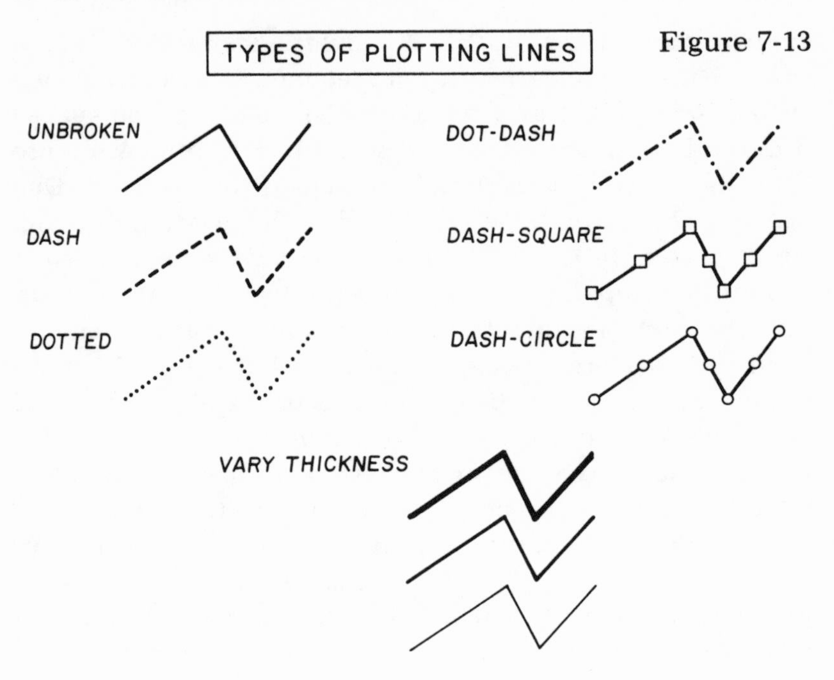

Figure 7-14

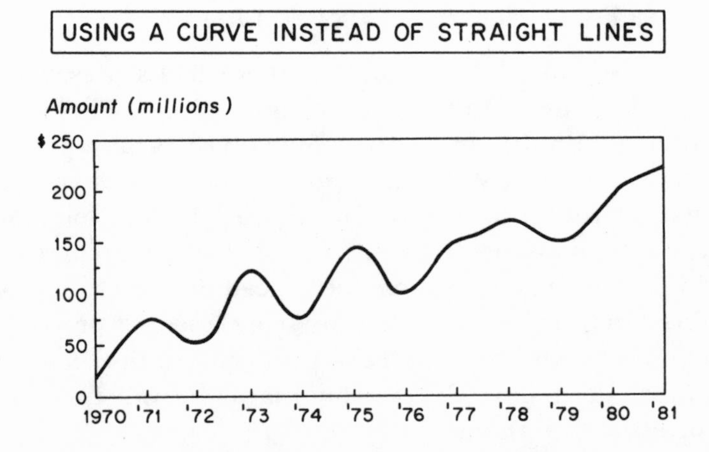

view, curves should be used only in certain special cases.)

Sometimes you may be working with data that have a direct linear relationship—that is, as a quantity of the amount on one scale goes up (or down), the quantity of the amount on the other scale also goes up (or down) at a similar rate. In these cases your plotted line will not be zigzag but straight. This is referred to as a straight-line graph, as in Figure 7-15.

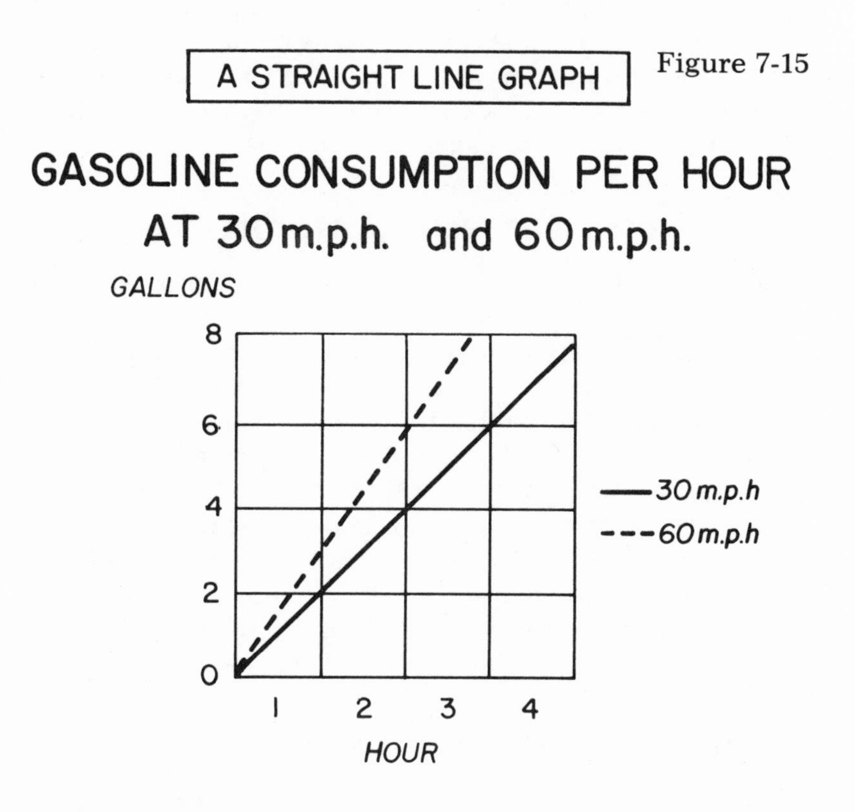

Figure 7-15

The plotted lines may be put in by using a felt or nylon-tip pen. It is also possible to purchase rolls of self-adhering pressure-sensitive black as well as many patterned tapes. These tapes come in suitably thin widths to use for plotted lines as well as other lines in the chart. (See Chapter 12, "Resources.") Remember that the plotted line or lines should be heavier than the other lines in the chart.

STEP FIVE: Transfer the Chart to Regular Paper

The final step is to redo the chart on regular unlined white typing paper. Most typing paper is transparent enough that you can place it over the draft chart and see all the lines, which can then be traced using a ruler. You can also measure the dimensions of your draft chart and redraw the chart on the white paper. You may find it helpful to cut out the draft along the four lines that mark its boundaries and place this on the plain paper and trace along the edges and mark off the points in the scales. Always be sure the chart is aligned parallel to the edges of your page.

MAKING REFERENCE TO THE CHART

When discussing a line chart in the text of your report you may wish to refer to the plotted line as the "curve" regardless of whether you use a jagged line or a curved line. This is appropriate since, in statistics, the term "curve" is used to refer to any kind of plotted line. You may also wish to refer to charts in which the horizontal scale represents some time period as a "time series chart." When the horizontal scale represents a category of measurement other than time (such as age, amount, rate, scores, size, or some other attribute) you may also refer to such charts as "frequency distribution" charts. You may wish to substitute the word "graph" for "chart" in referring to any kind of line chart.

OMITTING UNNECESSARY AREA

If the data are such that the plotted line does not approach the zero base of the vertical scale, you may wish to find a way

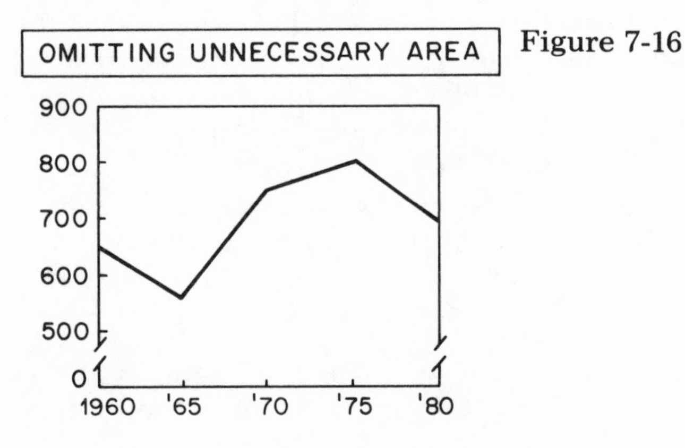

Figure 7-16

to avoid showing a good deal of space that has nothing plotted on it. In such a case you can start the scale with an amount higher than zero. Another method is to show the zero, but omit the unnecessary area by breaking the chart with a broken line, as in Figure 7-16.

USE OF SHADING WITH LINE GRAPHS

One of the more interesting variations of the line graph is to shade the area below the plotted line, as in Examples A and B, Figure 7-17. This is called a surface graph, since it calls attention to the area below the plotted line as a way of emphasizing the plotted line itself. The shaded area is prepared by cutting a piece of a shaded contact sheet to size, removing the backing, and affixing the sheet to the chart. Use your draft chart as a guide to cutting the contact sheet. You can also put in the shading by hand. When shading a line graph use hatching of various intensities or dots; do not use solid black shading, since this interferes with the reader's ability to follow the points in the horizontal scale.

SHADING THE LINE CHART

Figure 7-17

A. SURFACE GRAPH

OUTPATIENT
CARE EPISODE RATES, 1955-1975

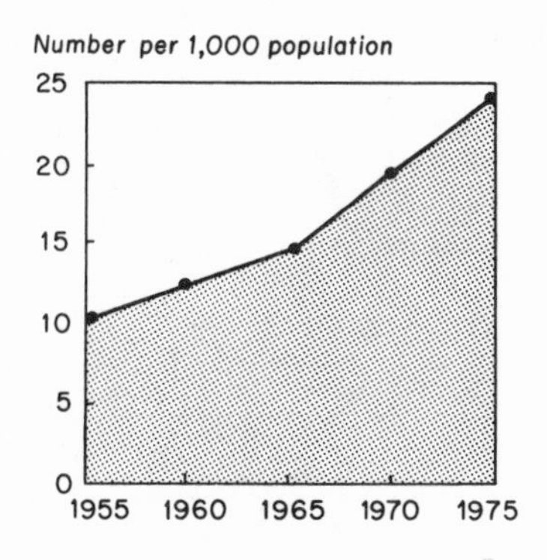

B. BAND OR STRATA CHART

INPATIENT & OUTPATIENT
CARE EPISODE RATES, 1955-1975

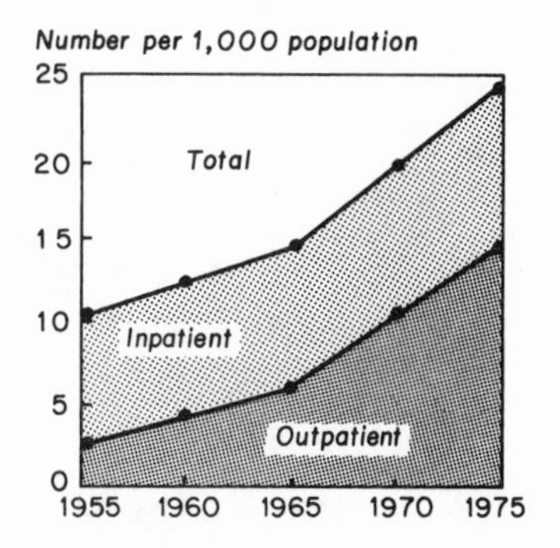

If the chart has more than one plotted line, then more than one type of shading can be used. This makes a band or strata chart, illustrated in Example B of Figure 7-17. Shaded charts of this type are sometimes incorrectly referred to as "area charts." You should avoid this term in your narrative, since it has a statistical use related to the measurement of functional relationships and variance.

USING PLOTTING MARKS

In order to prepare the plotted line it is necessary to locate plotting marks on the chart, as pointed out in Step Four. Once the line is drawn, however, it is optional whether or not to show the plotting marks or eliminate them and just use the plotted line. You will note that in earlier examples in this chapter some charts are shown with and some without the plotting marks to enable you to compare these techniques.

One advantage of showing the plotting marks is that the reader is helped to see more exactly the amounts being represented by the plotted lines. In cases of charts that might require many plotted points, however, the use of the marks can be distracting.

Figure 7-18

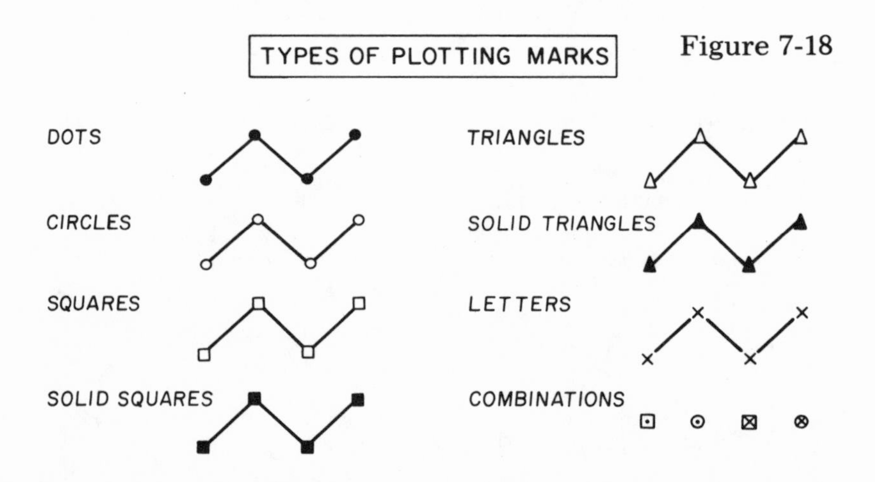

In charts with more than one plotted line it sometimes adds interest to the chart to show a different type of plotting mark for each different line. This method also helps to differentiate the lines. It can be further extended by using a legend that serves the purpose of identifying each plotted line without having to label the line. A number of different kinds of plotting marks are shown in Figure 7-18. There is also a template called Plotting Marks that can be used for this purpose.

SHOWING NEGATIVE VALUES

Occasionally the data upon which the chart is based include negative or minus values. In order to accommodate the line chart to these situations it is necessary to continue the vertical scale below the zero point. The layout for a chart with negative values is provided in Figure 7-19.

SHOWING NEGATIVE VALUES Figure 7-19

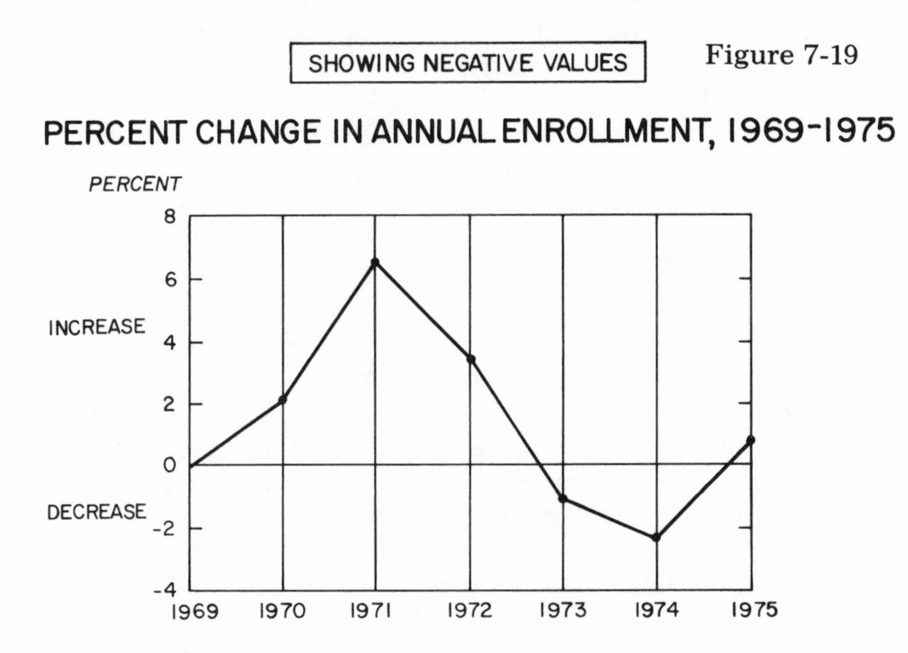

LABELING THE LINE CHART

Many examples of how to label line charts are shown in the charts in this chapter. As with all other charts, each line chart should have a chart number and title and these should be placed above the chart whenever possible. The vertical scale, which expresses an amount or percent, should be labeled with a heading that indicates the unit of measurement, such as "hundreds," "thousands," "millions," "percent," etc. The units in the vertical scale may be labeled on just the left side, or on both the left and right sides of the chart. Each unit of measurement that is marked off on the vertical scale should be labeled starting with 0 at the lower left-hand corner and proceeding upwards in equal steps, such, as 0, 100, 200, 300, 400, 500.

The units of the horizontal scale should be labeled along the bottom beginning with 0 and then proceeding with equal steps from left to right, such as, Jan., Feb., Mar., Apr., etc.; or 1960, 1961, 1962, etc.; or age groups such as 0–4, 5–9, 10–14, etc.; or scores such as 0–9, 10–19, 20–29, etc. Place a heading below the horizontal scale indicating months, years, age, score, or other units of measurement. It is permissible to omit the horizontal heading if the units are self-explanatory. For example, units designated as 1940, 1945, 1950, 1955 do not really require the label "years." The vertical-amount scale, however, should always have a heading.

STEP CHARTS

There are some very useful variations on the line chart. One, which combines it with some elements of the bar chart, is called the step chart. In this chart, illustrated in Figure 7-20, the plotted line is shown as a series of steps and gives the appearance of a staircase. This effect is achieved by drawing the plotted line as a series of straight vertical and horizontal lines rather than the angled or zigzag line that characterizes the usual line chart. In effect, the step chart is similar to the histogram or basic joined bar chart described in Chapter 5. The difference is that the step chart does not show the full vertical lines necessary to turn each step into a bar chart. The vertical and the horizontal scale of the step chart may also be ruled within the chart; the bar chart does not include ruled scales. The horizontal ruled lines are included in Figure 7-20, but can be omitted. If you put in the ruled lines be sure the step line is drawn heavier in order to distinguish it from the rules.

AN EXAMPLE OF A STEP CHART

Figure 7-20

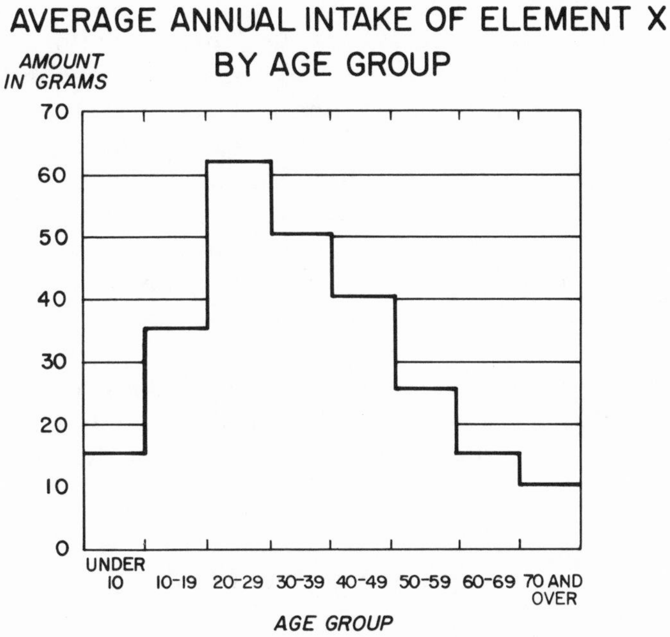

AVERAGE ANNUAL INTAKE OF ELEMENT X BY AGE GROUP

AMOUNT IN GRAMS

70
60
50
40
30
20
10
0

UNDER 10
10-19
20-29
30-39
40-49
50-59
60-69
70 AND OVER

AGE GROUP

The step chart is best used to plot only one factor, since the inclusion of more than one plotted line can be confusing to the reader. If you want to plot more than one factor, it is better to use a regular line chart.

The step chart is prepared by first locating each plotting point and then drawing a horizontal line through it for each unit on the horizontal scale. Each horizontal line is then connected with a vertical line.

The use of the steplike line gives the visual implication that the change from each horizontal scale unit (for example, age group or time periods) is quite sharp or up and down. The use of the angled or sloped line of the usual line chart implies a more gradual change. Step charts are especially useful to show averages, population figures, financial data, and data that tend to change sharply from one period to another.

HIGH–LOW CHARTS

Another specialized variation of the line chart is the high–low graph, which enables you to depict variations within each of a series of time periods. This is illustrated in Figure 7-21. It is often used to report changes in stock or bond prices, but can also be used to illustrate the range of prices, temperatures, or other factors for which you have data giving variation in a high and a low measurement during a given time period; or for any other factor shown on the horizontal scale. High-low charts do not show the average measurement for the time period, only the high and the low. For example, during the month of January the price of eggs may vary from 80¢ to 90¢ a dozen. The high-low chart enables you to depict this graphically. The average price of eggs may have been 84¢, but the high-low method will not show this. You can,

HIGH-LOW CHART

Figure 7-21

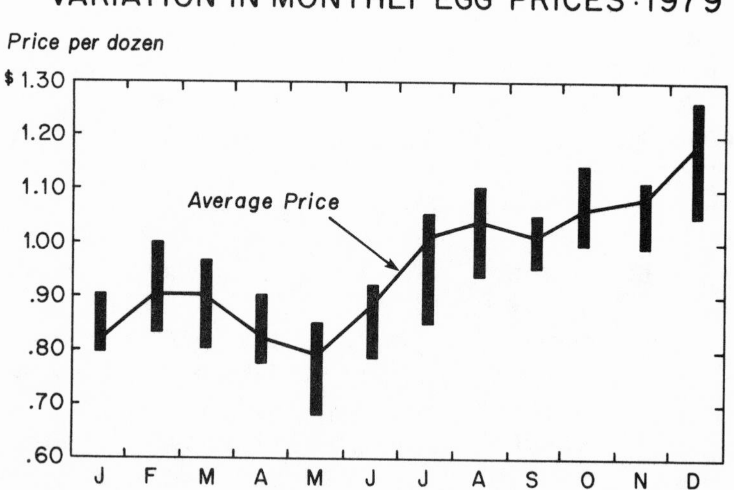

however, add a plotted line to the high-low chart to show the average price as well, as in Figure 7-21. Be sure, if you do this, to accurately label the line so the reader will have no doubt about what is represented.

The high–low chart, sometimes called a range graph, combines aspects of a bar chart and a line graph and represents an interesting example of how one can devise various combinations of different graphic methods.

CUMULATIVE LINE CHARTS

Figure 7-22 shows a cumulative line chart. The cumulative chart plots, for each unit in the horizontal scale, what the total amount is for that unit plus all the prior amounts. Thus, in Figure 7-22, savings deposits for a bank started in 1950 totaled $50 billion in 1955, and $75 billion in 1960. This $75 billion is the total net amount of savings amounts remaining in the bank from 1950 to 1960. By 1975 the bank had a total amount of $240 billion in savings accounts. A cumulative chart should always be labeled as such. Otherwise, it can easily be misunderstood. For example, without proper labeling it could be assumed that the $75 billion for 1960 might represent the amount deposited in savings accounts just for that single year rather than the cumulative amount.

Figure 7-22

CUMULATIVE LINE CHART
SHOWS RUNNING TOTAL

SAVINGS ACCOUNT DEPOSITS : 1950-1975

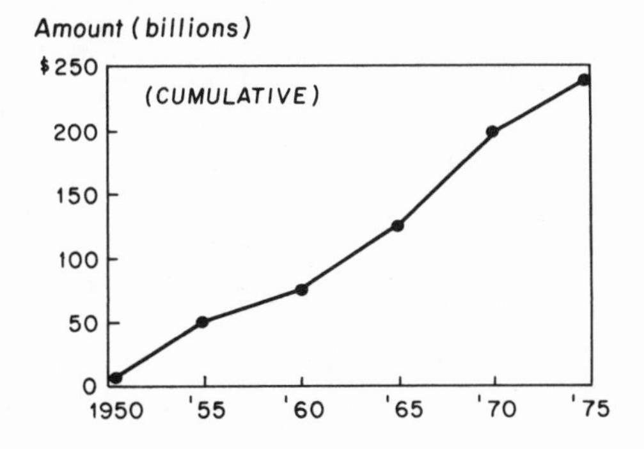

USING A NONQUANTITATIVE AMOUNT SCALE

Thus far we have referred to the vertical measurement scale as showing amounts expressed in quantitative terms. It is possible, however, to label the amount scale in nonquantitative terms or groups such as "high," "medium," "low." In Figure 7-23 a series of numerical ratings have been reclassified into nonquantitative groups. This simplification of the data can add interest and emphasis to the points you want to make in the report. However, one should always tell the reader what these simplified groupings represent in the narrative. For example, if the grouping "outstanding" represents scores of 90–100, this should be noted. In preparing the chart, of course, it will be necessary to use the original numerical data to locate the plotting points.

Figure 7-23

LABELING WITH NONQUANTITATIVE MEASUREMENTS

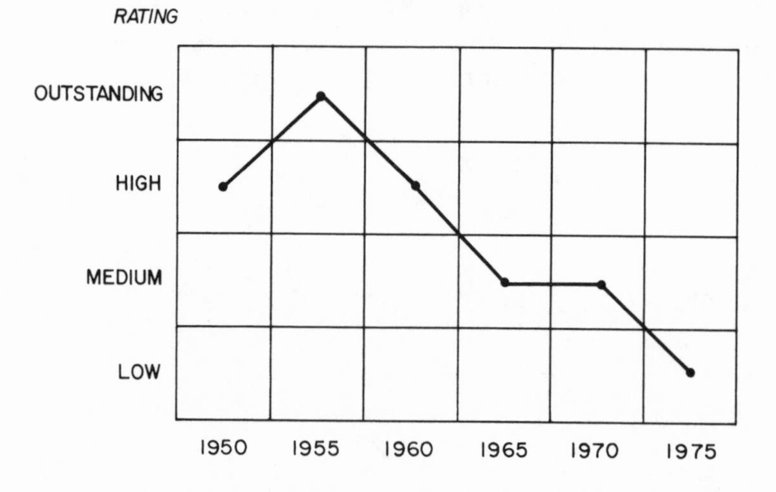

ALTERNATIVE SCALING AND SHADING

While we have stressed the convention of placing the amount scale on the vertical axis on the left side of the chart, there are times when you may want to vary this by placing the scale on the right. Scaling on the right is shown in Figure 7-24. It is sometimes desirable when you want to stress the data on the right side of the chart or when the plotted line goes up sharply. Another alternative is to show the scale on both the right and the left vertical axes. In Figure 7-24 we also illustrate some variations in shading to give a line chart additional interest.

Bar charts, pie charts, and line charts are the principal methods to depict quantitative information. There are also a number of graphic methods that are used exclusively for representing qualitative information. These are described in the following chapters on organization charts, flow charts, and time charts.

Figure 7-24

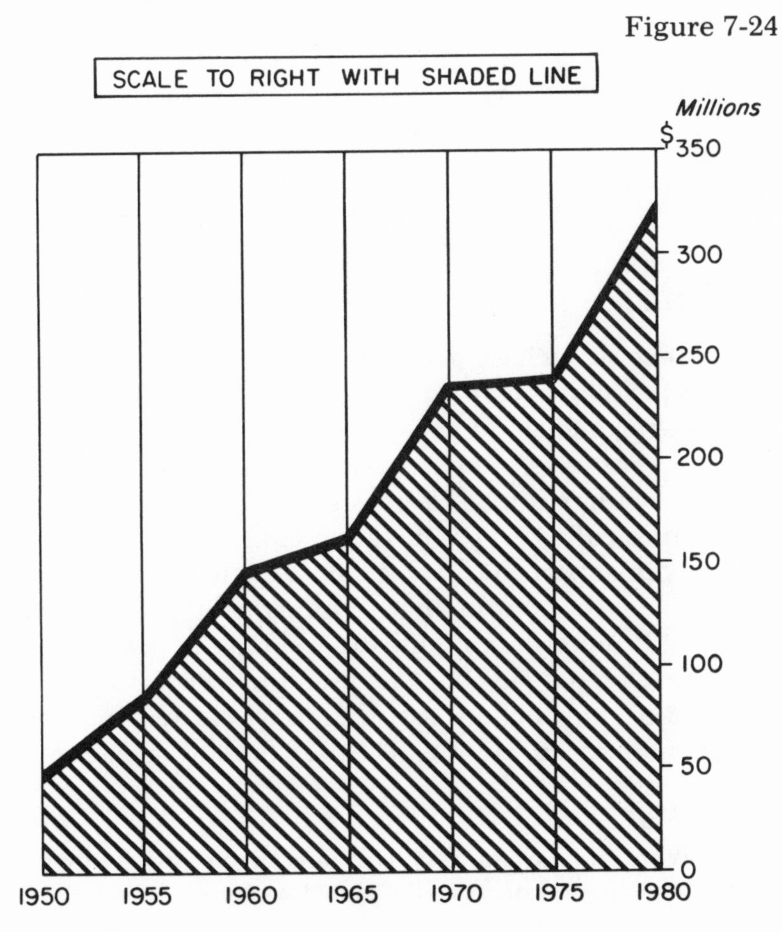

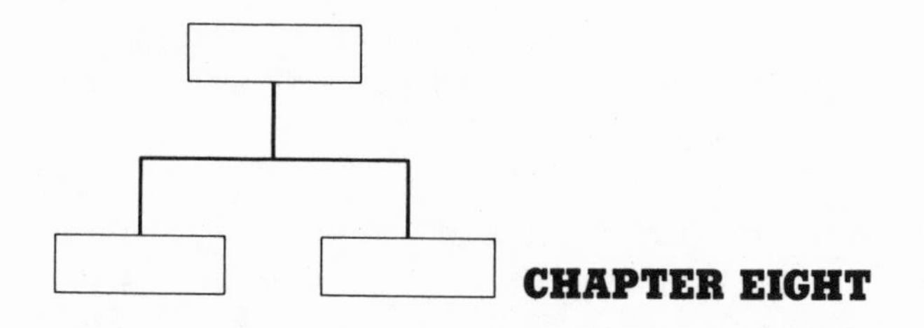

CHAPTER EIGHT

Organization Charts

WHAT IS AN ORGANIZATION CHART?

Many reports include a description of how organizations such as businesses, government, nonprofit agencies, institutions, departments, bureaus, associations, projects, groups, and other organized activities are structured. Such organizational structures can be effectively depicted by using a form of graphics called an organization chart.

An organization chart is a graphic device that uses pictorial methods to show qualitative information about an organization. It represents a quite different use of graphics than the bar chart, the pie chart, and the line chart, all of which are used to show quantitative or numerical information.

The organization chart can be used to show one or more of three things:

- What the various *staff* positions in the organization are, how they are structurally related to each other, and the span of control and chain of command within the organization.

- What the different *units* of the organization are and how they are arranged and related to each other.
- What the various *functions* are within the organization and how they are organized and related.

In order to prepare effective organizational charts it is imperative to be clear about the distinctions among the organization of staff, units, and functions. *Staff organization* refers to the manner in which the actual staff positions as designated by their titles are arranged. In addition to the position title, one may also show the name of the person occupying the position if this is relevant to the narrative discussion. Examples of position titles include Director, Associate Director, Personnel Director, Unit Chief, Bookkeeper, Manager, Dean, Salesman, Representative, Editor, Case Worker, Nurse, Engineer, and the like. This type of chart is also called a personnel chart.

Organization of units, on the other hand, refers to how the various formal subdivisions of an organization are structured. Each should be designated on the chart by the title of the unit. Examples of unit designations include Public-Relations Department, Research Division, Office of Special Services, Health Services Administration, Finance Office, Personnel Section, Board of Directors, Long-range Planning Committee, and similar designations. This is the most typical type of organization chart.

The *organization of functions* refers to the manner in which different activities, duties, or responsibilities for aspects of the organization's operations are related to each other. Examples of functions include public information, health services, planning, marketing, intake, production, data processing, etc. These may also be called function charts.

Sometimes these three aspects of organizational structure may correspond closely. The functional activity of providing public information, for example, may be assigned to a Public-Relations Director who heads a Public-Relations Department. But even if there is a high degree of correspondence it is important that the chart be internally consistent in what it is showing. A frequent error that is made in preparing organization charts is the random mixing together in the same chart of these three aspects of organizational structure.

These three uses of organization charts are illustrated in Figures 8-1, 8-2, and 8-3.

Figure 8-1

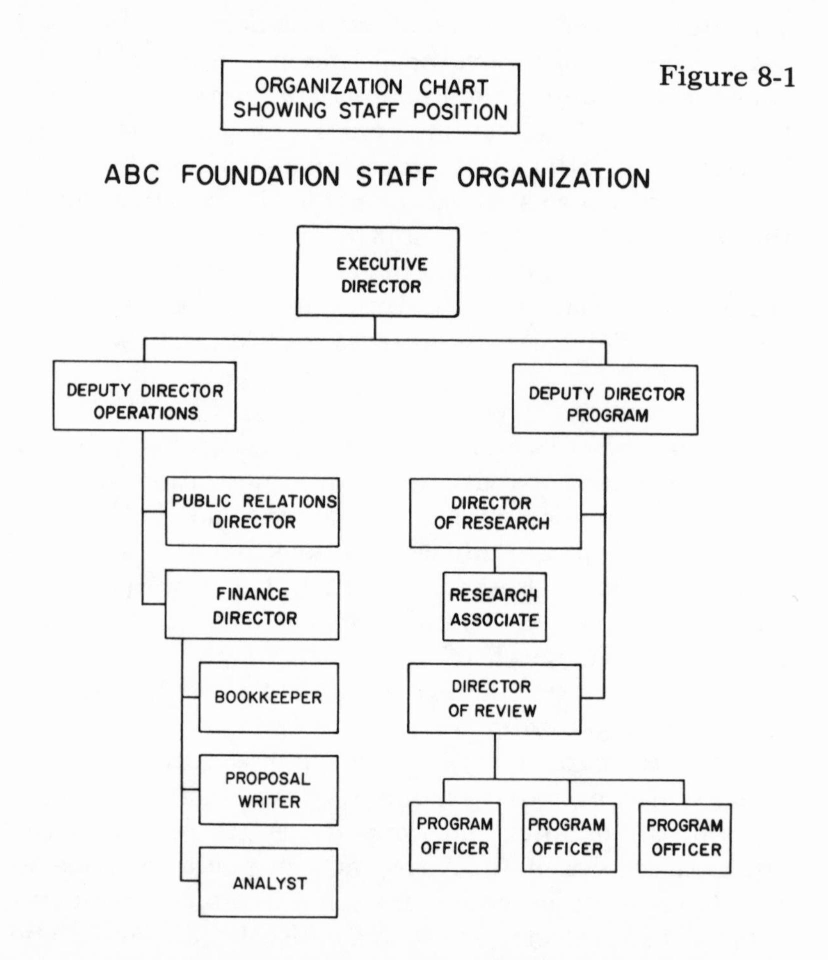

WHAT DO ORGANIZATION CHARTS ACHIEVE?

This type of chart is a very efficient way to describe an organization, since it can eliminate the need for a good deal of lengthy narrative description. The reader is able to grasp, at almost a single glance, the general structure of the organization being described. The chart can also convey a degree of

Figure 8-2

ORGANIZATION CHART SHOWING UNITS

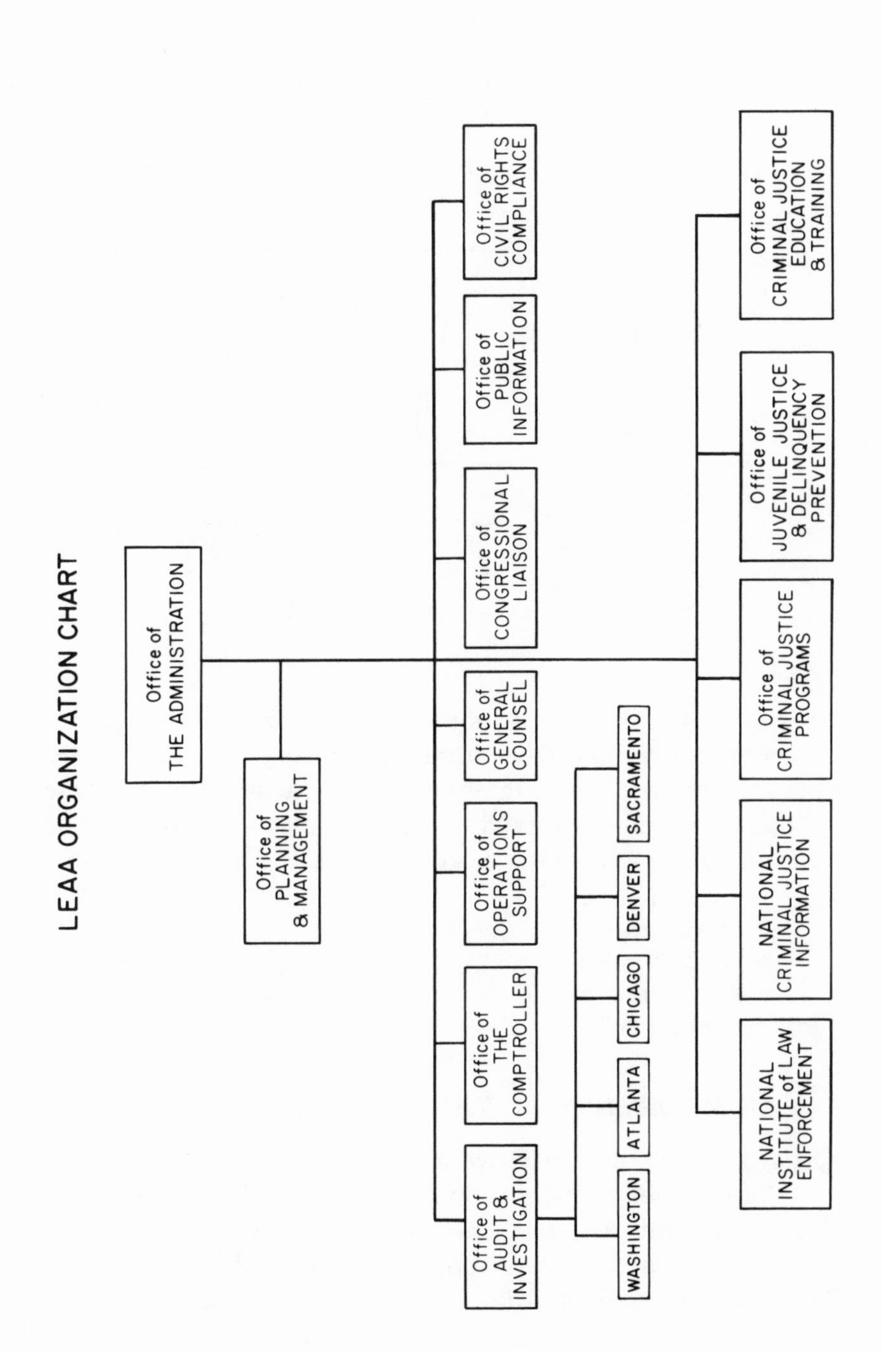

Figure 8-3

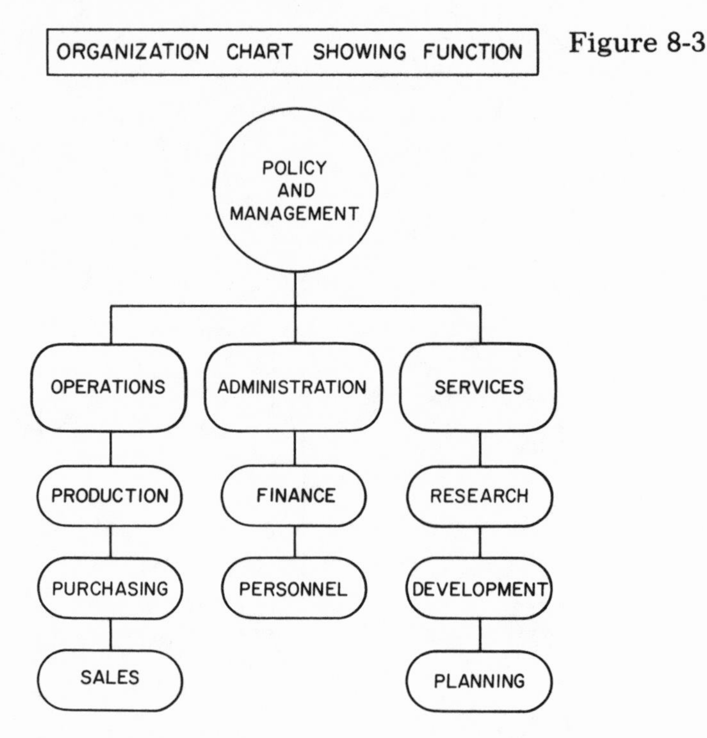

preciseness in the description of an organization that cannot be obtained by narrative alone. In addition, the inclusion of this kind of chart in your report is regarded by the reader as a sign of a well-organized and systematic operation. Thus, organization charts effectively achieve the objectives of:

- clarification
- simplification
- summarization
- credibility

COMPONENTS OF THE ORGANIZATION CHART

Organization charts use two basic graphic elements: (1) rectangles or other shapes, and (2) lines that are solid or dotted. These elements are combined in graphic presentations that enable the reader to picture the structure of the organization by the use of a box or other shape to represent each position, unit, or function of the organization. The relationships among the positions or units or functions within the organization are represented by the connecting lines. The overall effect is such that the chart conveys a sense of the extent to which the organization's structure represents a coordinated whole.

Organization charts are comprised of four components, which include:

- The title and number of the chart
- Rectangles or other shapes that represent staff positions, organizational units, or different functions
- Labels that designate what each rectangle or shape represents
- Connecting lines that indicate the relationships among the positions, units, or functions that are represented

HOW TO PREPARE AN ORGANIZATION CHART

Organization charts generally show a rectangle or oblong box for each position, unit, or function being described. It is also possible to use circles, triangles, and other shapes to add interest or to represent different organizational levels or different types of positions, units, or functions. For convenience we shall usually refer to this range of shapes as "boxes" in the following discussion. There are six steps involved in preparing these charts, as follows:

STEP ONE: Plan the Chart

The first step is to plan the chart. Most organization charts require one-half to a whole page in a typewritten report. If you use a whole page, leave margins of no less than 1½″ on each side. As with other kinds of charts, using a separate page for the chart will make the typing of the report as well as the preparation of the chart much easier.

The main decision to be made in the planning stage is to determine the number of different boxes or other shapes to be included in the chart. Jot down on a worksheet a list of all of the positions, units, or functions of the organization that you want to account for in the chart. The minimum number of boxes is three. The maximum number that will enable you to label them clearly on a full page (assuming you will type the labels) is about twenty.

Since most organizational structures are hierarchical and resemble a pyramid, there are usually more boxes at the bottom of the chart than at the top. Therefore, organization charts that will show many units are frequently placed sideways on the page in order to leave more room for the larger number of units at the bottom of the chart. As with other charts, you should attempt to avoid the necessity for the reader to turn the report sideways, since this is an impediment to maintaining reader interest.

If your list of units is too large to be accommodated on a single page, find ways to combine smaller units into larger components. Having done this, you can prepare an additional chart to show the more detailed breakdown of those units if you want to describe them more fully. For example, your organization may have a Community-Relations Department with four area offices. If, on your overall chart, you have to show many other departments and do not have room for the area offices, use one box for the Community-Relations Department. On a second chart show the detail of the way that department is organized. The decision regarding the extent to which you want to add additional charts, of course, depends on whether they are necessary to enhance the narrative description or the points you want to stress in your report.

Step Two: Draft the Chart Layout

There are two major choices with respect to the layout of the boxes. One is to lay them out in a way that maximizes the horizontal relationships of units; the other is to emphasize vertical relationships. Figure 8-4 illustrates these two approaches for the same organizational structure. It should be noted that maximizing vertical relationships usually increases the number of boxes you can get on the page and label without difficulty. It also makes it more feasible to show the chart on the page lengthwise so that the reader does not have to turn the report sideways to read the chart. More often than not, the structure of organizations requires that the two approaches be combined, as in Figure 8-1 referred to earlier. However, you will find that it is still necessary to decide which layout (vertical or horizontal) is to be emphasized more.

Figure 8-4

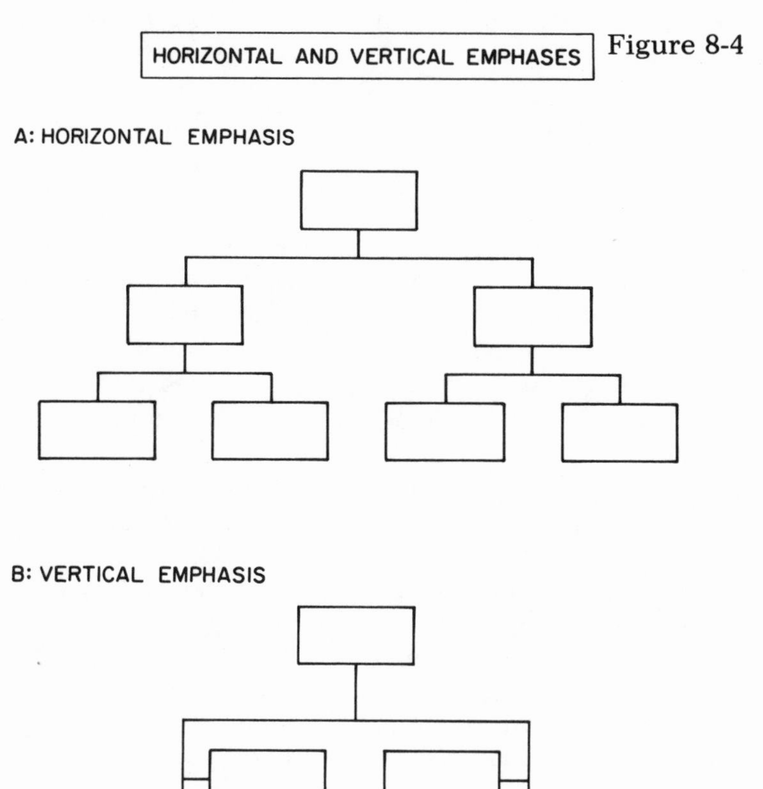

STEP THREE: Decide on the Size of Boxes and Labels

It is not feasible to establish any rules regarding the size of the boxes or other shapes you may be using. Obviously, if you only have five boxes to go on a page, they can be larger than if you have to fit ten boxes on a page. In part, your decision will also be determined by the labeling of the boxes. If you want your label to go inside the box (this is generally preferable), the box must be large enough to accommodate the label. On the other hand, sometimes the number of boxes you need to show may preclude the possibility of having them all large enough to fit the label inside the box. In such cases you can label next to the box.

Step Four: Position and Label the Boxes

Positioning the boxes or other shapes used in an organization chart always requires some experimenting. This can best be done in draft form using a worksheet of ruled graph paper. The ruled lines of the graph paper enable you with only a ruler to draft boxes of accurate size and with straight lines. There are also special templates just for this type of chart; the templates are comprised of rectangles in a variety of sizes. You can also make use of the symbols on the computer template.

Using graph paper will aid you to obtain uniformity in the distances between the various boxes in the chart. Start with the boxes at the top of the chart and work your way down. If you seem to be running out of space, you might experiment with (a) smaller boxes, (b) changing the placement of the boxes, or (c) combining boxes.

As part of this step, write in the proper labeling for each box. Remember that a typed eight-letter word such as "Director" will require about ¾″. (Depending on the type style being used, this will vary by a fraction of an inch.) Be sure your boxes are large enough to accommodate the label. You should leave no less than two typewriter spaces of blank space between the label and the sides of the box. One trick in doing this is to have all the necessary labels typed separately and cut each one out to use in preparing the boxes.

Another trick is to prepare all of the boxes (or other shapes) with the typed label and cut out each box. Move the boxes around on the page until you have them lined up just where you want them and glue them to the page with rubber cement.

STEP FIVE: Draw in the Connecting Lines

Each box should be connected to at least one other box with a line. Two kinds of lines are most frequently used in organization charts. A solid line (—) represents a direct relationship, line of authority, or chain of command within the organization. A broken line or dashes (- - -) represents an important relationship, but not a direct relationship. Solid lines are used to show that a position or unit reports administratively to another unit or position. Broken lines may show indirect reporting relationships or lines of communication within the organization.

STEP SIX: Transfer the Chart

After you have all the components of the chart laid out in draft form, the next step is to transfer the chart to regular typing paper. There are different techniques for doing this. One is to redraw the chart completely, using your draft as a guide to measure off the boxes with a ruler or by using the boxes of the template. Another method is to prepare each box with its typed label, cut it out, and glue it on your final sheet with rubber cement. Then draw in the lines with a ruler or edge of the template. This method is very satisfactory if the report is to be photocopied, since the photocopied page will not reveal that each box has been glued on.

BALANCE

One of the important qualities that can make or break the effectiveness of an organization chart is the extent to which it appears to be well balanced. Naturally, lines must be perfectly straight and should be parallel to the sides or the top and the bottom of the page. Avoid angled lines if at all possible.

In addition, you should place the boxes in such a way that the left and the right sides of the chart appear to have more or less equal weight. While exact balance may not be possible because of the nature of the organizational structure itself, you should try to work in this direction. It may help to use boxes of different sizes. For example, there may be just a few boxes on the left side and many on the right. Use smaller boxes on the right side so that the total amount of space occupied on each side of the page is similar.

You must, however, be careful in selecting box sizes. To some extent the size of the box can imply the relative importance of the person or the unit in the organization. Thus, the box used for a department head should not usually be larger than the box for a senior vice-president. In most organizations there are more people in the lower ranks and, therefore, it is necessary to use smaller boxes as you go down the chart.

There are, of course, organizational patterns that do not correspond to the typical pyramidal hierarchy. Some administrative, staff, committee, or production units, for example, may be organized on a more collective, peer, or coequal basis. When this is true be sure you reflect such a structure graphically by using boxes of equal size and placing them on a parallel level in the chart.

VARIATIONS ON THE ORGANIZATION CHART

One of the more interesting ways to vary the basic organization chart is to use a variety of shapes such as circles, ellipses, and triangles in addition to rectangles or squares. This can make the chart more visually appealing and increase its impact. It can also enable you to show more information in the chart than if you use just boxes. Different shapes can be used to convey different administrative and organizational levels, functions, or responsibilities. When you use a variety of shapes it is sometimes necessary also to include a legend to make clear what the symbols represent.

In the next chapter we will describe flow charts, another popular type of chart used to portray qualitative information other than organizational structure. Many of the shapes that are described for use in flow charts can also be adapted for use in organization charts.

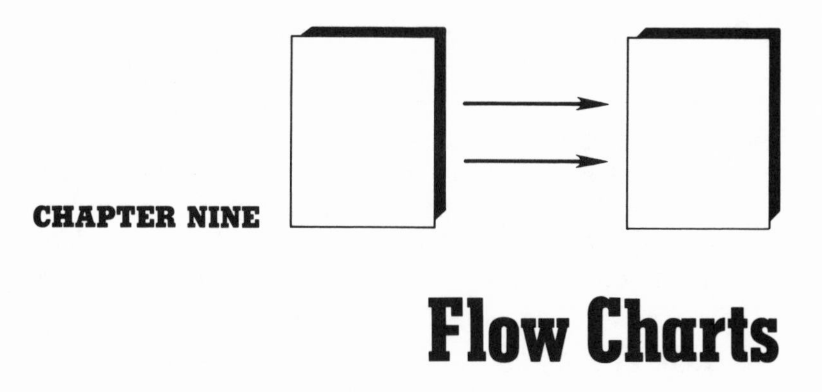

CHAPTER NINE

Flow Charts

WHAT IS A FLOW CHART?

A flow chart is a graphic method to show pictorially how a series of activities, procedures, operations, events, ideas, or other factors are related to each other. It shows the sequence, cycle, or *flow* of these factors and how they are connected in a series of steps from beginning to end. Like the organization chart, the flow chart is used to depict qualitative information rather than the quantitative relationships depicted by bar, pie, and line charts.

Flow charts, when used in connection with industrial engineering, systems analysis, and computer information systems, can be extremely complex and technical. We are concerned here, however, with the more simplified forms of this method that can be employed to enhance written reports of a less technically specialized nature.

Flow charts of this kind are most useful to achieve:

- ▶ clarity
- ▶ simplification
- ▶ summarization
- ▶ coherence

Flow charts can condense long and detailed descriptions of a series of activities into a single chart that contributes to simplifying and making such descriptions more clear and concise. The flow chart also conveys the sense that a series of steps or ideas have a systematic and coordinated relationship to each other and thus adds coherence and unity to the presentation. In addition, use of the flow chart can add to the credibility of the material represented by it in the report.

COMPONENTS OF THE FLOW CHART

Flow charts have four basic components:

- A title and chart number
- Labels for each element shown in the chart
- Shapes such as boxes, triangles, circles, diamonds, and others generally found on most templates used for computer diagramming
- Lines or arrows that connect the shapes and show the flow from one to the next

Flow charts may be prepared by using a ruler or a template such as the computer diagrammer template. If you use a variety of different shapes such as those on the computer template, you may include a legend that designates what each shape represents. Figure 9-1 provides you with a guide to some of the typical shapes that can be found on most templates and how they can be used.

You can, of course, make up your own shapes, or you can use just a single shape, such as a rectangle, for the entire chart. It enhances the sophistication of your report, however, to utilize the variety of standard computer template symbols even though the material represented in the chart may be unrelated to computer systems.

The components of the flow chart may be diagrammed so that they run in straight lines in either a horizontal direction, a vertical direction, or both, on the page. They can also run in a circular direction. Samples of these various formats are shown in the examples in this chapter. Many flow charts are comprised only of boxes and arrows, but it will add to their interest and effectiveness if you can use the variety of shapes and symbols shown in Figure 9-1.

Figure 9-1

TYPICAL FLOW CHART SYMBOLS AND SHAPES

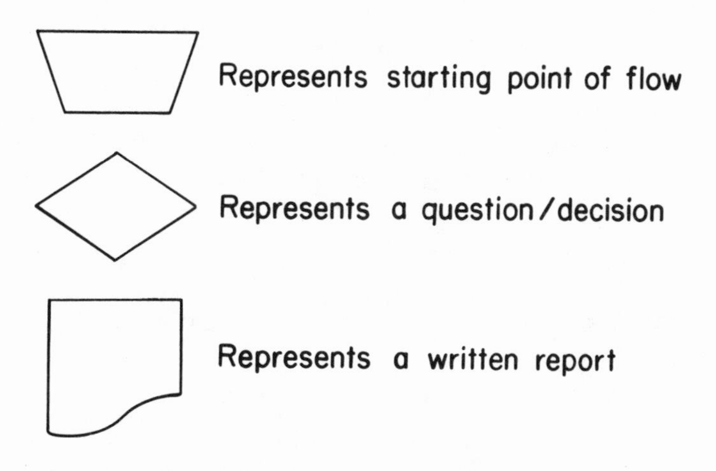

These represent activities:

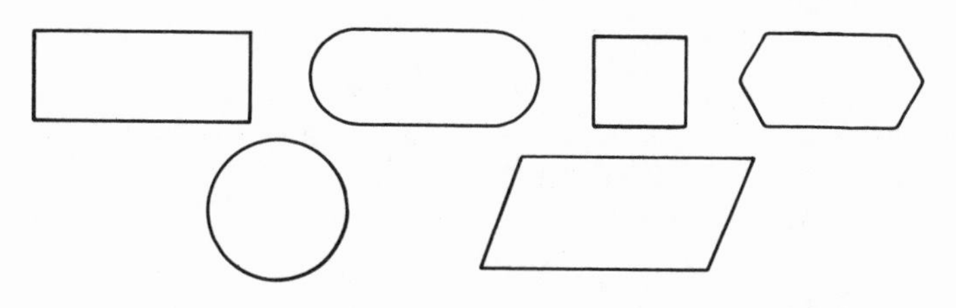

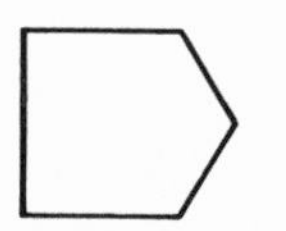

Off-page connector to show chart is continued to the next page

Show direction of flow with arrows

Many different formats are possible with flow charts. Preparing flow charts can be a very creative and satisfying activity since you are not bound by the more rigid rules that apply to some of the graphic forms described in earlier chapters but may use your ingenuity. Keep your charts as simple as possible and avoid the frequently made mistake of being tempted to show everything in the flow chart. Basically, what the flow chart does is to stress a set of major activities or ideas and their sequential relationships.

HOW TO PREPARE A FLOW CHART

There are a number of general steps and principles that should be followed in preparing a flow chart even though, as pointed out previously, there can be great variation in the form these charts take. These steps and principles include the following:

STEP ONE: Decide on the Appropriateness of the Use of a Flow Chart

A flow chart cannot be effectively used in every case where your report is describing activities or qualitative relationships. The most typical cases in which the flow chart may be appropriate are when your report describes one or more of the following:

- A series of specific *procedures, activities,* or *tasks* that are followed, such as the steps involved in an application procedure shown in Figure 9-2
- The movement of *persons* through a particular process, as shown in Figure 9-3
- Causal connections, linkages, or relationships among various ideas, concepts, or constructs, as shown in the input-output model in Figure 9-4

It will help you to decide on the appropriateness of a flow chart if you become familiar with the variety of types of these charts, so spend some time studying the examples.

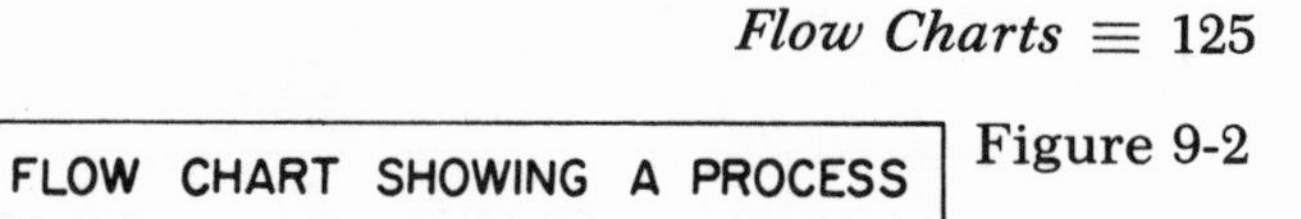
FLOW CHART SHOWING A PROCESS

Figure 9-2

LICENSE APPLICATION PROCESS

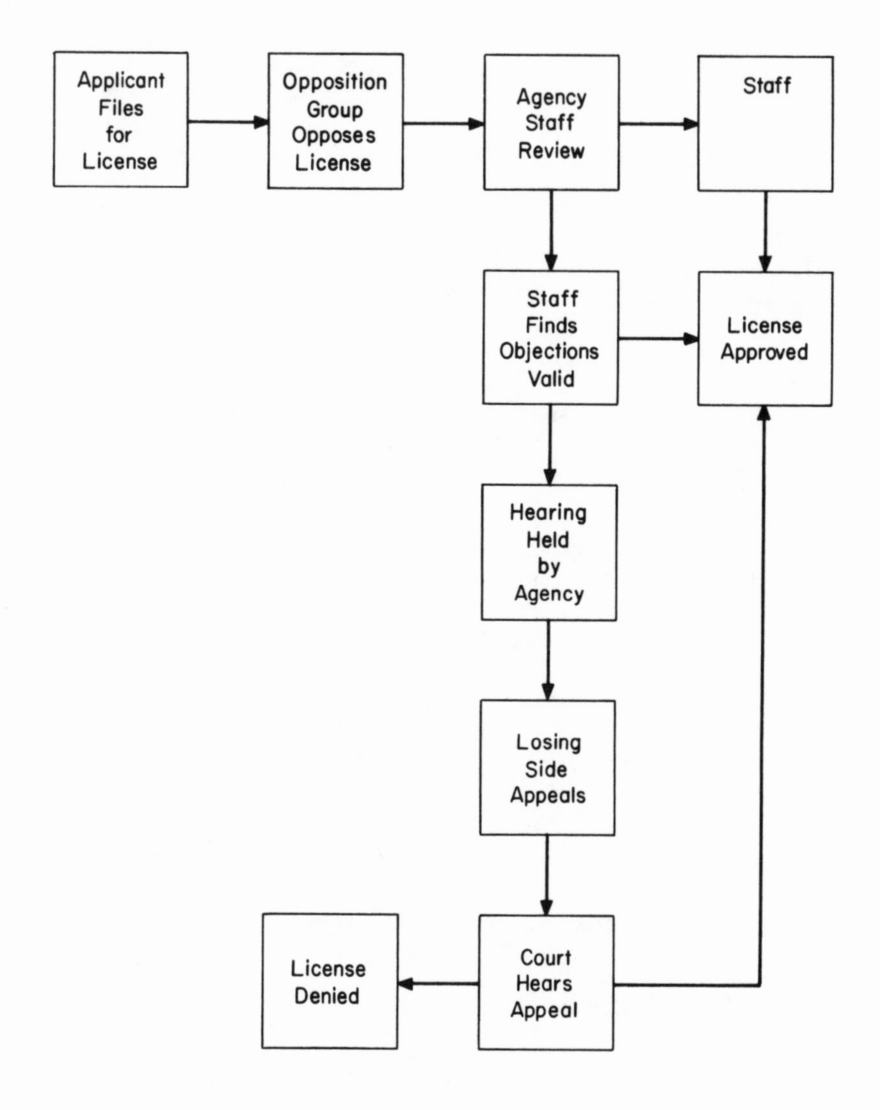
Applicant Files for License
Opposition Group Opposes License
Agency Staff Review
Staff
Staff Finds Objections Valid
License Approved
Hearing Held by Agency
Losing Side Appeals
Court Hears Appeal
License Denied

Figure 9-3

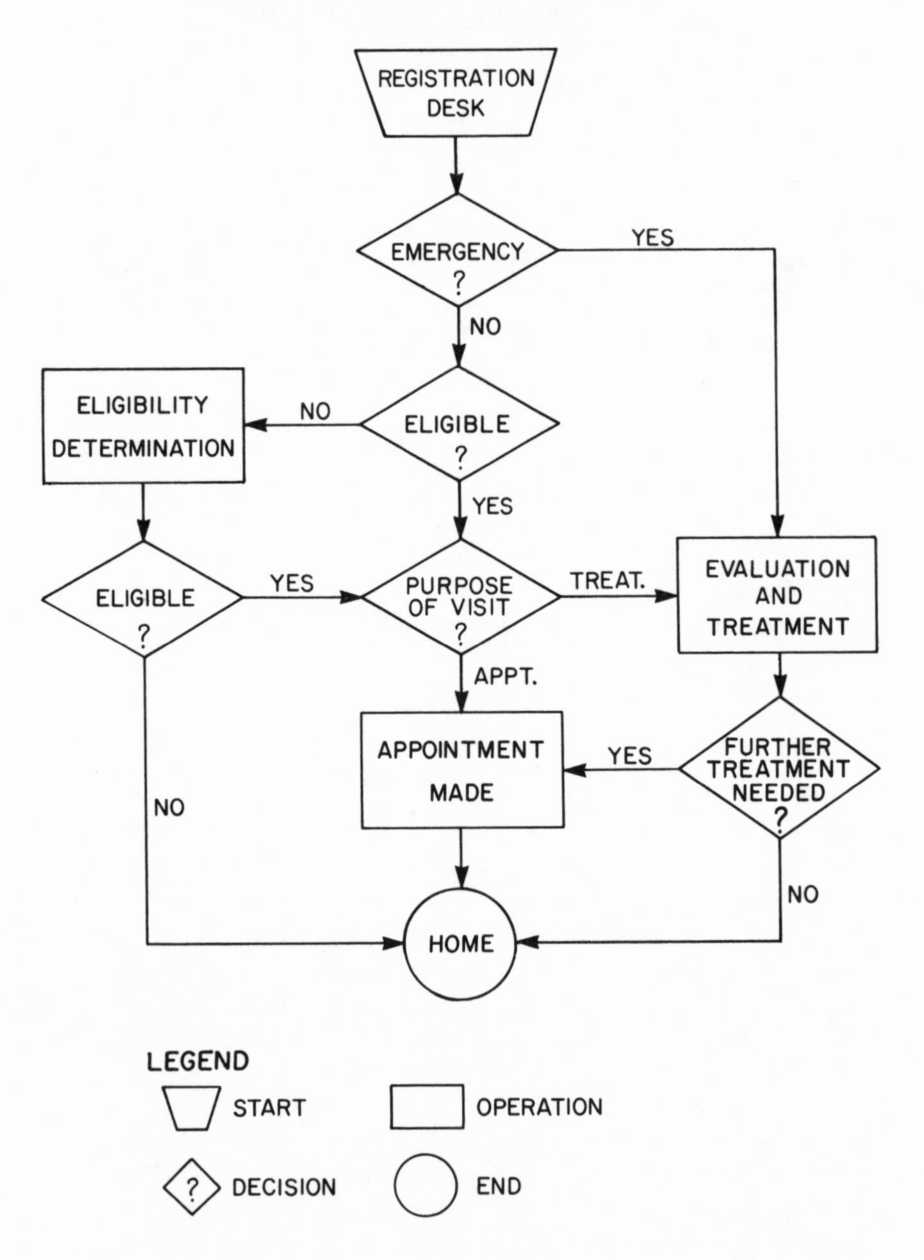
FLOW OF PERSONS
HEALTH CENTER FLOW CHART
REGISTRATION DESK
EMERGENCY ?
YES
NO
ELIGIBILITY DETERMINATION
NO
ELIGIBLE ?
YES
ELIGIBLE ?
YES
PURPOSE OF VISIT ?
TREAT.
EVALUATION AND TREATMENT
APPT.
APPOINTMENT MADE
YES
FURTHER TREATMENT NEEDED ?
NO
NO
HOME
LEGEND
START
OPERATION
DECISION
END

Figure 9-4

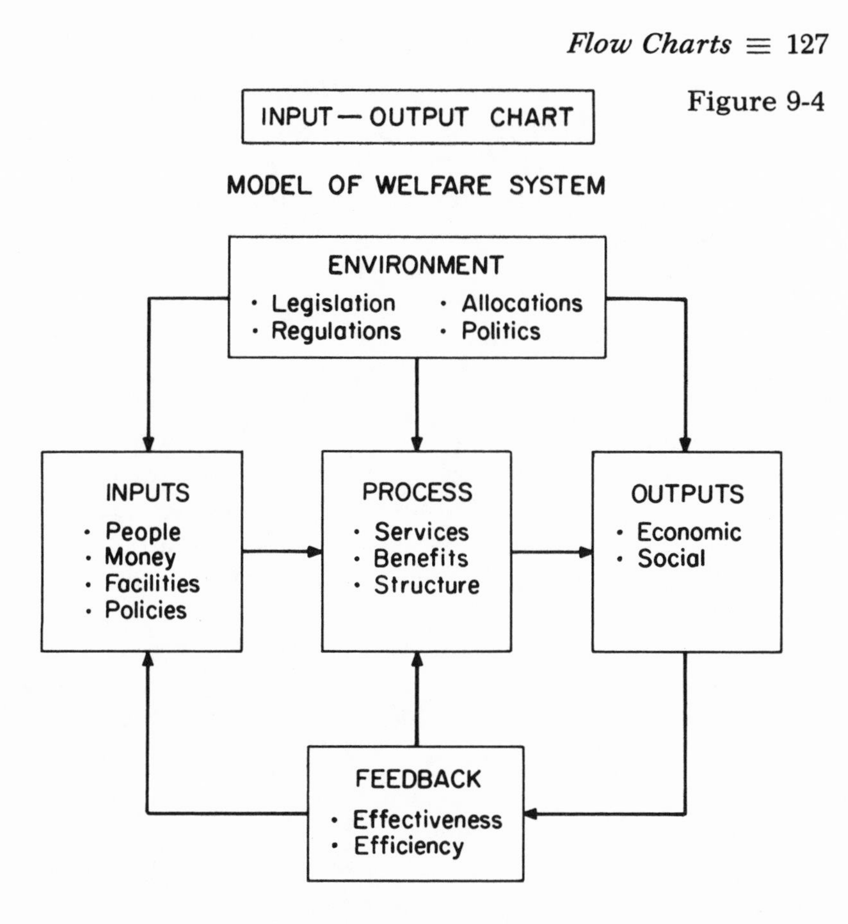

STEP TWO: List the Items to Be Included in the Chart

Once you have decided to use a flow chart, the next step is to decide on the items to show in the chart. You can do this by listing (a) the components (that is, the major activities or ideas) that you want to include in the chart and (b) the elements (subactivities or ideas), if any, that comprise each component. Think in terms of components and elements or of major activities and minor activities. For example, in the input-output chart shown in Figure 9-4 the components of this chart are inputs, process, outputs, environment, and feedback. Each of these is represented by a large box in the chart. The elements are the subdivisions such as the items listed under Inputs, which are people, money, facilities, and policies. An alternative way to draw this chart could have been to show each element in a smaller box flowing into the input component. It should be noted that whenever a separate box is used for an item, it is given more prominence in the visual effect, with the resulting implication that it is of increased importance in the total scheme being shown.

Sometimes you will be using the flow chart to describe situations in which the various activities are all pretty much of the same importance (as in Figure 9-2) and, in these cases, the idea of components and elements does not apply.

In preparing the list of the individual items you want to include in the chart, list them in sequential order and note their relationship to each other. While you may seem to be limiting yourself to linear thinking when you do this listing, this is not necessarily true. By using multiple arrows and by constructing circular and various other layouts, you can reflect nonlinear relationships, as in Figure 9-5.

Figure 9-5

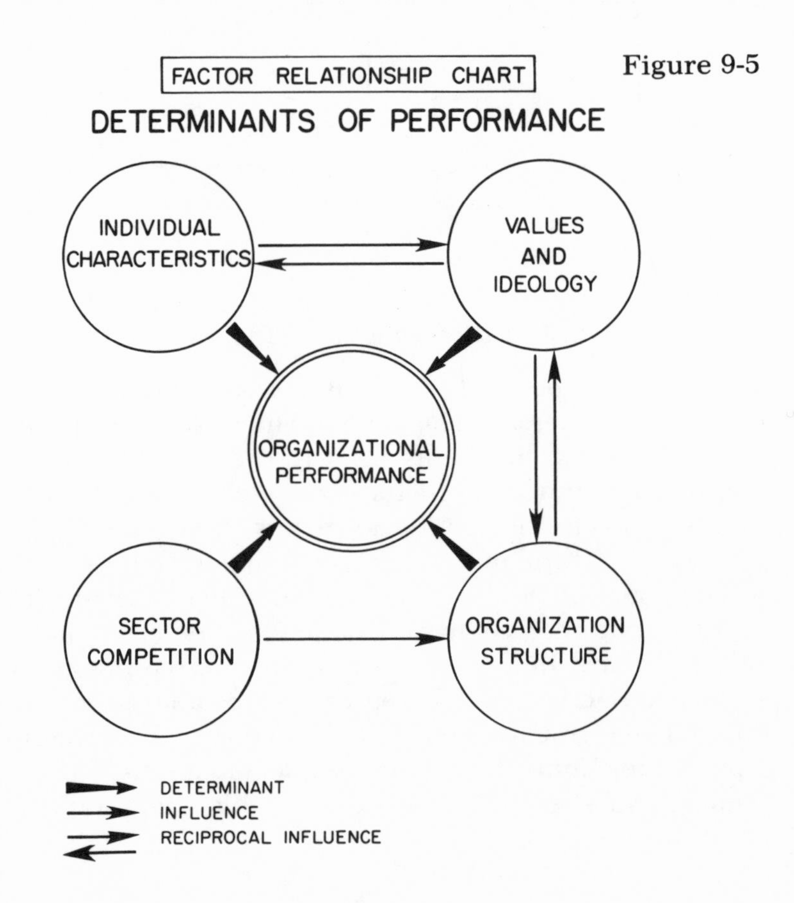

Arrows and lines are used to represent the direct relationships. This means that you must decide on what the major relationships are in order to know where to place these lines. In a simple description of the step-by-step sequence of a process, as in Figure 9-2, this is relatively easy to do. But in presentations of more conceptual material (for example, Figure 9-5), such decisions are more complex.

STEP THREE: Sketch Alternative Layouts

Most flow charts require you to do a certain amount of experimentation in order to decide on the final layout. You must decide on what kinds of geometric shapes to use, what they are to represent, where to place them, how they are related, and how to label them. Since there are no hard-and-fast rules to guide these choices, it is expedient to prepare a number of preliminary rough sketches of possible designs in order to arrive at a final selection. Use ruled graph paper for these sketches. This will help you to locate the placement of the various shapes and to keep the chart balanced and in proportion.

As you work on these sketches to refine the chart you may find it necessary to modify, add, or eliminate items from the list prepared in Step Two.

STEP FOUR: Draft Final Design

Once a design has been chosen you are ready to prepare the chart in final form. The easiest way to do this is to draft the flow chart on lined graph paper and then recopy it on regular paper.

USING FLOW CHARTS FOR INSTRUCTION

In recent years the flow chart has been widely used as a tool in presenting instructional and training materials. The central principle in developing such material is to identify each step or operation that is required in carrying out an activity. These steps must be charted in sequence. The outcome of each step must be reduced to specific alternative next steps.

The conditions under which one can go to any alternative next step must be specified. All of this information is then charted using shapes to show the steps; lines and words to designate conditions for moving to the next step or steps; shapes to designate the next step, and so forth. In Figure 9-6 we see an example of one way to present such instructional material.

Figure 9-6

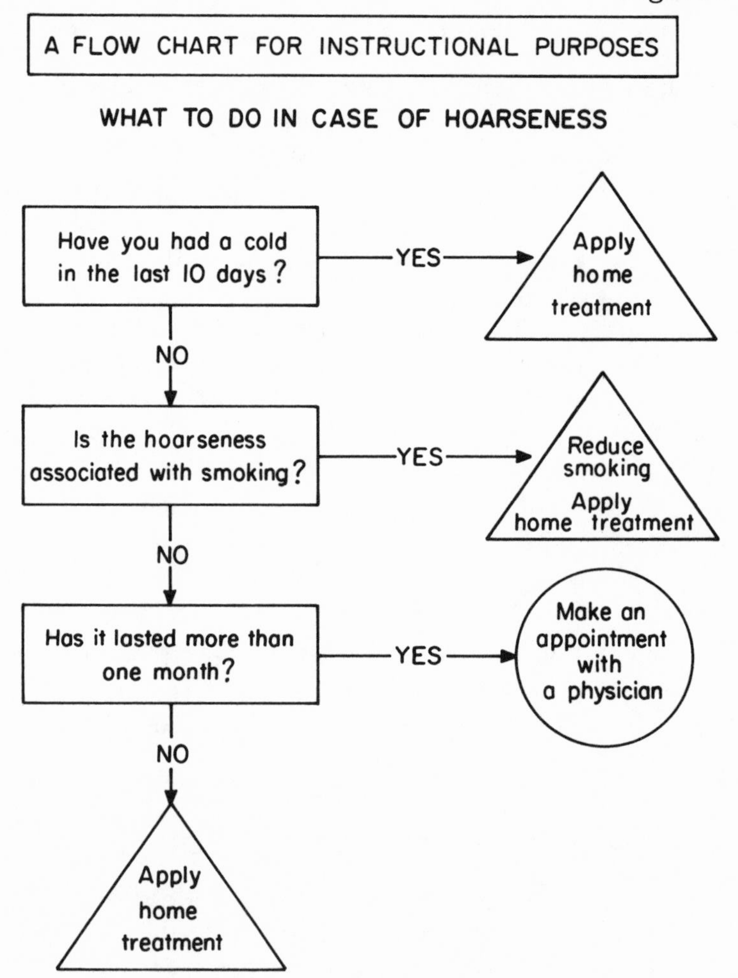

Reprinted from *Take Care of Yourself* by Donald M. Vickery, M.D., and James F. Fries, M.D. Copyright © 1976, by permission of Addison-Wesley Publishing Co., Reading, Mass.

CHARTING THE RELATIONSHIP OF FACTORS

Explaining the manner in which various factors are related to each other is one of the frequent ideas that are included in reports prepared by government and organization officials and by various professionals. The basic flow chart can be adapted for this purpose.

Such ideas may have to do with (a) what factors appear to contribute to the creation of some particular problem or situation; (b) what factors seem to influence the manner in which an organization performs; or (c) what factors tend to be the determinants of the behavior of individuals or groups of individuals. The explanations can be clarified and also made more convincing by supplementing the text with a chart that we will call a *factor-relationship chart.*

For example, in explaining the factors that influence the manner in which an organization performs, you might identify four major sets of factors. One set of factors are the characteristics of the individual or individuals who are the chief decision makers, such as their family backgrounds, their educational levels, and their prior organization experience. A second set of factors might be related to the manner in which the organization is organized, its resources, and its size. A third set of factors might be related to the economy of the specific sector in which the organization operates, including the intensity of competition. A fourth set of factors might be the particular ideology, beliefs, and values that characterize the organization. Figure 9-5 shows how a chart can be used to illustrate this analysis and, by viewing such a chart, the reader may be increasingly persuaded of the validity of the argument.

It is important in preparing factor-relationship charts to develop broad categories of factors, since the chart is always a simplification of more complex conceptual explanations. A shape such as a circle or a box is used to represent each category of ideas and is appropriately labeled. Arrows of different styles or in combination serve to show relationships and interactions among the categories. A legend, as in Figure 9-5, can be used to explain the meaning of the different types of arrows that are used.

The possible variations you can invent for using flow charts are almost endless. Combining several shapes and symbols that are accompanied by clear labeling can add a great deal of interest to complex narratives and, at the same time, bring additional clarity and coherence to the explanation. Flow charts may be used not only to enhance narrative reports but also as useful tools in the processes of management and planning.

A specific adaptation of the flow chart that has realized increased popularity is the time chart, which is described in the next chapter.

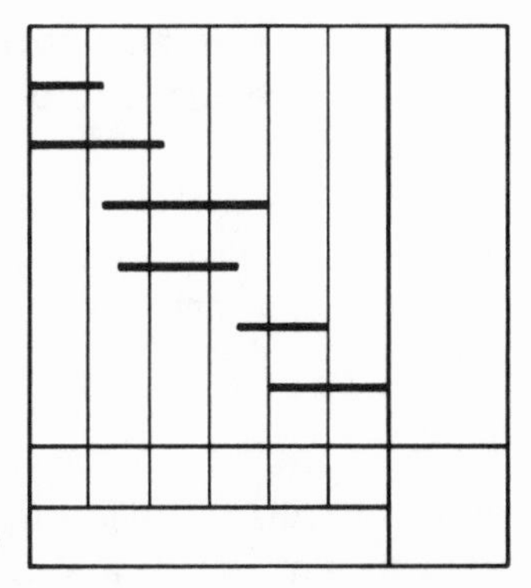

CHAPTER TEN

Time Charts

A time chart is a type of graphic that shows the timetable for carrying out a series of activities, procedures, or tasks. Many reports describe a schedule of activities that take place over a period of time. The time chart is a very efficient way to convey in a concise and clear manner a good deal of detailed information regarding what is often a complex schedule of activities. It is especially useful in describing the steps involved in planning and implementing all types of programs and services, establishing marketing and purchasing procedures, acquiring and opening new facilities, clarifying procedures for processing applications, setting production schedules, and for sequencing training programs, research activities, and the like. Time charts are especially important in the area of proposal preparation, since most governmental granting agencies and private foundations require or expect that a time chart will be included in a grant proposal.

A time chart includes a list of each activity and the time scheduled from its start to its completion. Use of the chart as a device to describe timetables achieves a number of objectives in a report, including:

- clarity
- summarization
- coherence
- credibility

The time chart has the additional virtue of providing a sense of coordination to a series of tasks, since all activities are shown within a single chart. This creates the visual impression that the activities are coherently related to each other and are coordinated.

The most frequently utilized kinds of time charts fall into two groups. One is known as Gantt, time-line, or milestone charts. These are the easiest kinds of charts to prepare and are very effective. The other group is referred to as PERT charts, a more complex and technical kind of chart.

The Gantt chart, named for the industrial engineer who developed it, shows the beginning and end for each of a series of activities or events. The time-line and milestone charts are variations of this technique.

In Figure 10-1 a typical Gantt chart is used to describe the timetable for a research interview study.

PERT charts, on the other hand, are much more complex ways of depicting time schedules. PERT stands for Program Evaluation Review Technique, which was originated by the Westinghouse Corporation during World War II as a way to schedule the many interrelated activities involved in the construction of submarines. PERT charts show the relationship among all the scheduled events and activities and how they are interdependent for completion of any entire program. In addition, PERT utilizes the Critical Path Method (CPM), which shows the maximum period of time that can be taken to complete each step in a total process and still finish within an overall time frame. Most PERT charts are very complex and require the use of the computer. For this reason they will be discussed here only in terms of some of the simplified adaptations that can fruitfully be made of this technique.

BASIC GANTT CHART

Figure 10-1

TIMETABLE FOR INTERVIEW STUDY

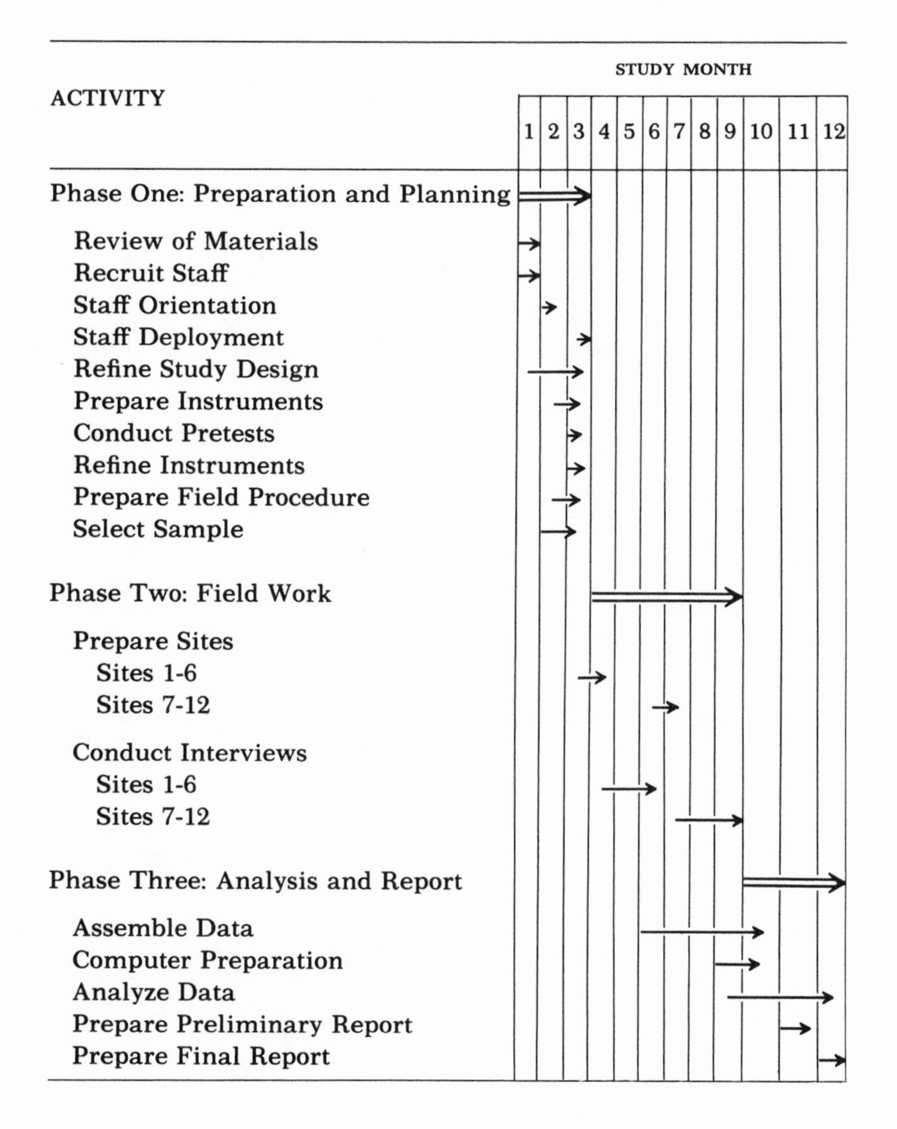

COMPONENTS OF A TIME CHART

A time chart has four basic components:

- A chart number and title
- A list of a series of events, activities, or processes
- A grid or series of markings that depict a time series expressed in minutes, hours, days, weeks, months, or years
- Lines, arrows, or symbols that show the amount of time elapsed for each activity from its beginning to its completion

The following is a step-by-step description of how to prepare the basic Gantt chart, with references to some of the variations that can be made.

HOW TO PREPARE A TIME CHART

There are four main steps involved in preparing a time chart, as follows:

STEP ONE: List and Group Activities

The initial step in preparing a time chart is to list each activity that you wish to describe. Use brief phrases of one to four or five words; otherwise, there will not be sufficient room on the chart. For example, in explaining a staff recruitment and training program, the narrative description of the first set of activities might be as follows:

> Job descriptions will be prepared for each position by reviewing existing official definitions in the agency's personnel manual, state manuals, the *Dictionary of Occupational Titles,* descriptions of similar positions in other agencies, and other sources. These materials will be analyzed by the staff and draft descriptions will be written and submitted to the Personnel Committee for review. Following this review, the descriptions will be revised and submitted for approval to the Director's office.

For purposes of planning the time chart this narrative should be converted to a list of activities, as follows:

Obtain materials
Review materials
Prepare draft descriptions
Have Personnel Committee review
Revise descriptions
Submit to Director's office

It will increase the effectiveness of time charts if a number of individual activities can be grouped under general headings. For example, the foregoing set of activities can all be listed under an overall heading, as follows:

Preparation of Job Descriptions
Obtain materials
Review materials
Prepare draft descriptions
Have Personnel Committee review
Revise descriptions
Submit to Director's office

In effect, what is being done by using this format is to show a major activity (preparation of job descriptions) and the subactivities (review of materials, etc.) that comprise the major activity. Additional major-activity headings in such a recruitment and training program might include: Recruitment; Staff Selection; Job Assignment; and Orientation and Training Activities. Each of these headings would have a set of subactivities under it. Each major heading may be further designated as a phase of the study, such as *Phase One: Preparation of Job Descriptions*; *Phase Two: Recruitment*, and so forth.

Step Two: Decide on Time Measures and Estimate Elapsed Time for Each Activity

Once all activities are listed and grouped, the next step is to estimate the time required to carry out each of the activities. This is best done using a worksheet on which you list each activity and the approximate amount of time required for its completion.

In order to do this, it is first necessary to decide on what unit will be used to express time. Should it be in years, months, weeks, or days? Deciding on the time-unit measurement depends on (1) how long the total project will take and (2) how much detail it is necessary to show. If a total project is to take a year, the time may be measured in units of weeks or months. If a project takes a total of a month or two, it is logical to use days or weeks as the measurement. In preparing the time estimate for each activity, do not hesitate to use fractions of whatever unit you decide on.

After you decide on the time required for each activity you can then estimate the time requirement for the phase or component. Using the above example, and assuming the entire project will take one year, the following is an example of this step:

Example of Time-estimate Worksheet

RECRUITMENT AND TRAINING PROGRAM—ONE YEAR

ACTIVITY	TIME REQUIRED
A. Preparation of Job Descriptions	1 month
1. Obtain materials	1 week
2. Review materials	1 week
3. Prepare draft descriptions	1 week
4. Personnel Committee review	1 day
5. Revise descriptions	1 week
6. Submit to Director's office	1 day
B. Recruitment	2 months

You will notice that the time estimates for the subactivities add up to more than the time for the major activity (phase) under which they fall. This is so because some of the activities would be going on concurrently during the one-month period.

STEP THREE: Prepare Chart Layout, Time Grid, and Headings

The third step is to prepare the chart layout. This is done by first putting in all the headings at the top of the chart; second, measuring off the distance to be used for the time periods (that is, the time grid); and last, drawing the time columns with vertical grid lines. You can do this very effectively by drafting the chart on lined graph paper, which will help you keep all the items in the chart evenly lined up and make it easier for the person who will later type the chart. Be sure that the width of the space for each time period is kept the same.

SAMPLE TIME DESIGNATIONS

Figure 10-2

Sample A:

Activity	Month								
	1	2	3	4	5	6	7	8	9

Sample B:

Activity	Month								
	Jan.	Feb.	Mar.	Apr.	May	June	July	Aug.	Sept.

Sample C:

Activity	Month								
	1	2	3	4	5	6	7	8	9

Sample D:

Activity	Month								
	1	2	3	4	5	6	7	8	9

The chart headings that designate time periods may be expressed in a variety of ways. Figure 10-2 includes some samples of the most frequently used modes of time-period designations. While on some occasions it may be necessary to indicate considerable detail, as in Examples C and D, it is advisable to keep the chart as simple as possible.

STEP FOUR: Enter Activity Lists and Time Lines

In this step the final entries are made on the chart. All activities should be listed down the left-hand side of the chart. If activities are grouped (for example, Phase One; Phase Two, etc.), include the group heading in this list and indent individual activities under the heading. This format is illustrated in Figure 10-1.

After this list is complete, put in the time lines in the form of an arrow that designates the beginning and the end of each activity. If you have a group heading that spans three months, as does Phase One in Figure 10-1, be sure that the activities listed under this heading cover the entire three-month period from beginning to end. If they do not, then either reduce the time period covered by the heading so that it is consistent with the total period covered by the individual items or extend the time for certain subactivities. If the items under the heading cover more than three months, extend the time for the heading. Use a heavier line or a double line for the time line of the major headings.

In the process of completing this step you will probably find that you are doing a good deal of planning and making revisions in time estimates. Thus preparation of time charts also becomes an important planning tool in addition to an effective graphic. Once you have all activities listed and the time lines entered on the chart, it is ready for final typing and reproduction.

There are many different ways to represent the activity time lines in addition to using arrows. Some of these are shown in Figure 10-3. It should be noted that when the lines include the use of milestones, squares, or similar symbols it is important to show the exact point where the activity ends. This is done by placing the symbol on the time line itself, not

Figure 10-3

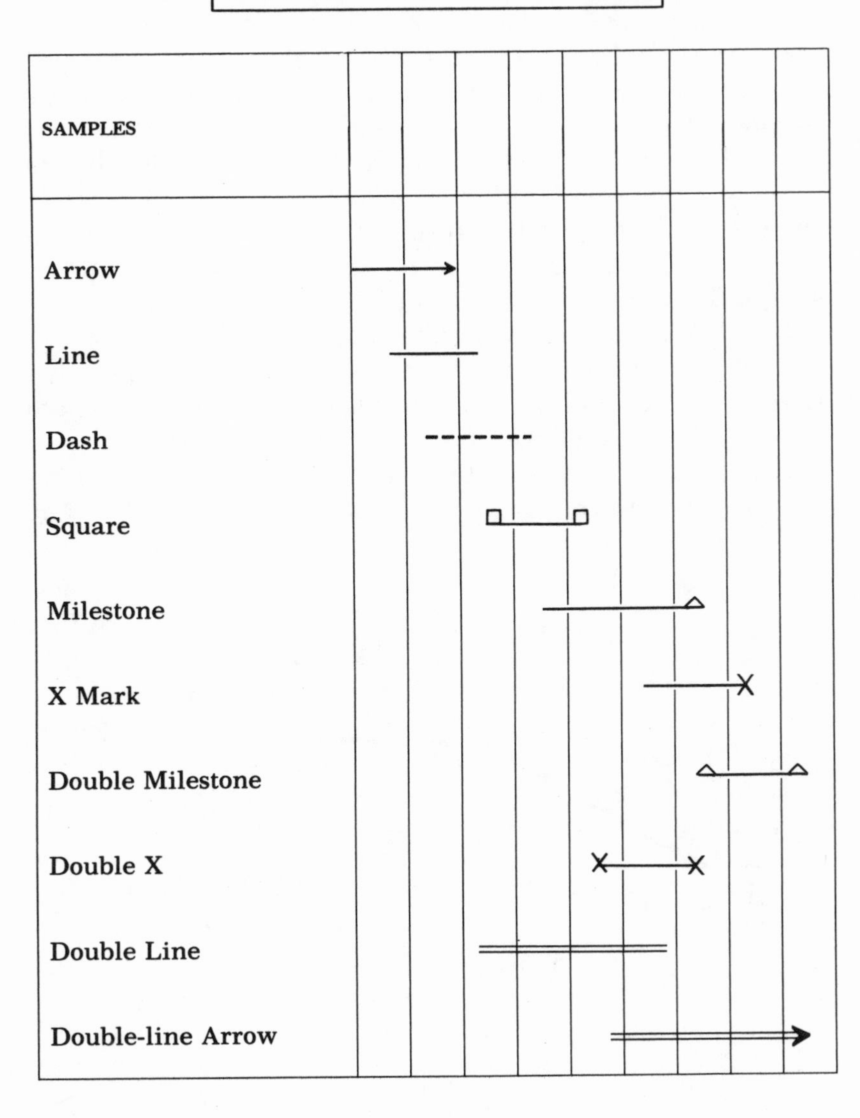

as an addition to the end of the line.

Time charts such as we have been discussing are sometimes referred to as *milestones,* because the symbol resembles the triangular stones used along New England roads to mark off distances in miles.

PERT CHARTS

PERT charts are a type of time chart that are much more complex than the Gantt, milestone, or similar charts. PERT is often a computer-assisted technique because of the large number of items that are represented. It is usually inappropriate for use as a method of graphic presentation in the kinds of reports with which this book is concerned, since in such reports it tends to confuse rather than clarify. On occasion it can be adapted for use, however, and such a chart, in sample form, is shown in Figure 10-4.

The PERT chart is comprised of a series of circles joined by lines. In this chart a circle represents an event. The line represents the activities that must take place for the event to happen, but does not tell you what those activities are. As in Figure 10-4, it is necessary to use a legend to explain what the event is that is represented by the numbered circle. The

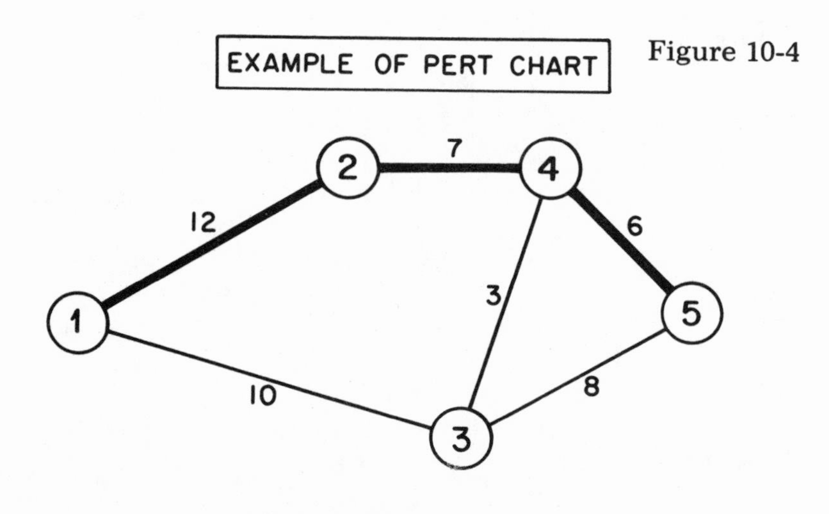

Figure 10-4

EVENTS:
1 PROJECT AUTHORIZATION
2 OFFICE SPACE ACQUISITION
3 STAFF HIRING
4 STAFF TRAINING
5 STAFF DEPLOYMENT

numbers on the lines represent the time elapsed, usually expressed in days. Thus, in Figure 10-4, it takes 12 days to go from event 1 (project authorization) to event 2 (office-space acquisition). The heavy line represents what is known as the "critical path," which is determined by the path that takes the most days from the start to the finish.

VARIATIONS ON THE TIME CHART

One of the variations of the time chart is that which combines it with a flow chart. An example of this variation is shown in Figure 10-5. The boxes indicate the sequence or flow of steps in an application process. Overlayed on the flow chart is a time frame that shows the amount of time required for each step in the process.

COMBINED FLOW AND TIME CHART — Figure 10-5

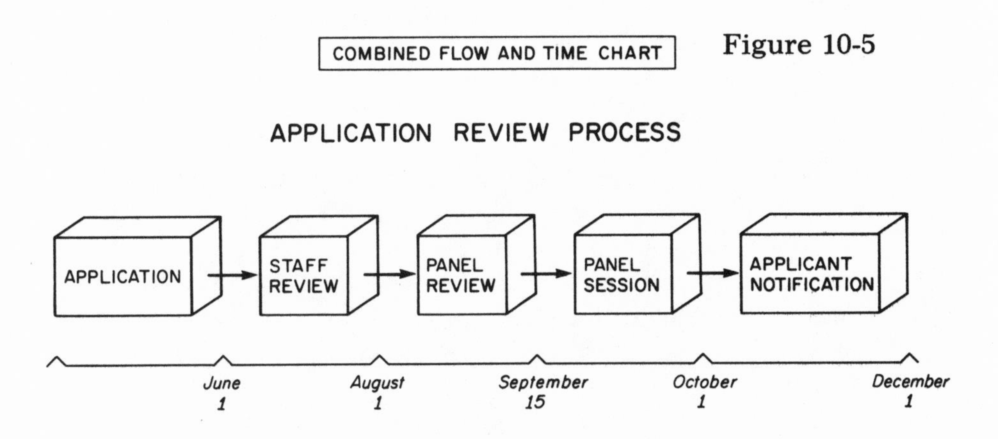

CHAPTER ELEVEN

Adding Interest and Emphasis

There are a number of easily executed devices that can be used to add interest and emphasis to narrative reports. These particular devices consist of the direct application of symbols to the layout of the narrative text itself. These devices are described and illustrated in this chapter. They include the use of symbols such as bullets, squares, lines, boxes, and shading.

There are a wide variety of symbols that can be used to add interest and emphasis, and many of these are shown in Figure 11-1. They are almost all available in a single template known as a Plotting Mark template or on transfer sheets, which many art and office-supply stores carry. In the following section we shall describe the use of the bullet as an example of how these kinds of symbols can be used.

USING BULLETS

The term "bullet" refers to the use of dots of various sizes as a technique to make a report more interesting. The standard

Figure 11-1

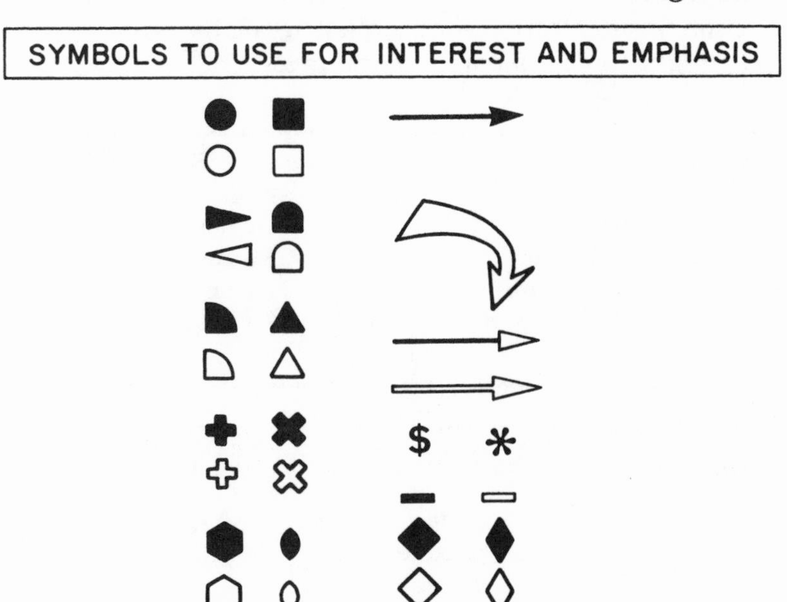

typewriter period or o can be used, or you can put the bullet in by hand with a pen. Use felt-tip, nylon-tip, or ball-point pens to make dots larger than the typewriter period. Making the dot darker than the typewritten text with a pen adds not only visual interest but also additional emphasis. Bullets are employed primarily in connection with presenting lists. Compare the two examples, A and B, below, and you can see how the bullets tend to emphasize the listing:

Example A: No bullets

design project
collect data
analyze data
prepare report

Example B: Single Bullets

- design project
- collect data
- analyze data
- prepare report

Bullets may be used individually, as in Example B, above; or in a consistent group, as in Example C; or in series, as in Example D.

You will note in Example D that the bullets serve a further function in substituting for numbering the list which, otherwise, might have appeared as follows:

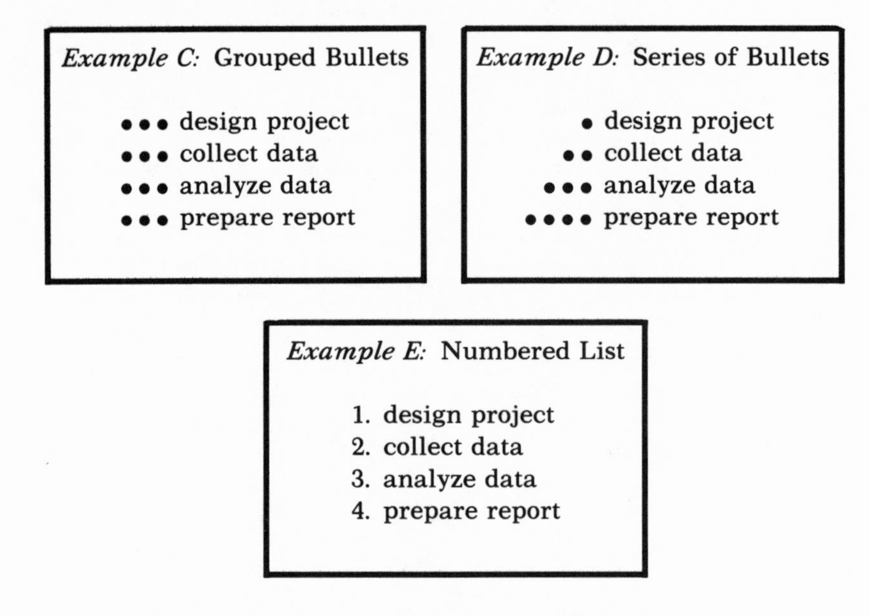

If in the text or in an anticipated discussion you believe you will need to refer to items in a list by their actual numbers, then you should not insert bullets as symbolic substitutes for numbers. Bullets should not be used in combination with numbers, since this can create confusion and lose the purpose of the bullet, as seen in Example F.

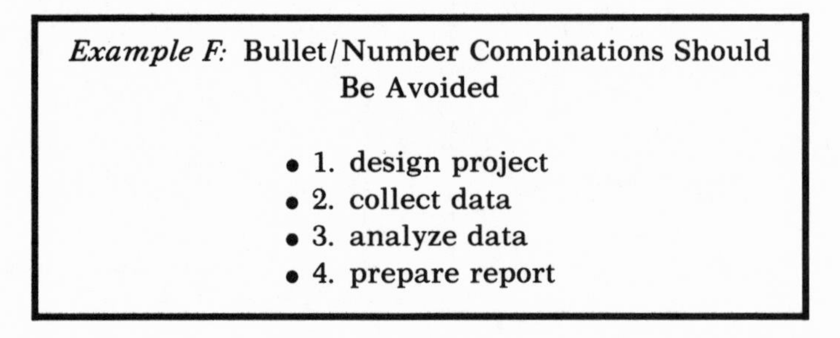

Bullets can be used with lists that are comprised of one or two words or very short phrases, as in the preceding examples, or they may also be used effectively with lists that are comprised of a series of longer paragraphs that you want to emphasize. An example of this would look like the following:

The reporting system is comprised of three principal elements:

- *Objectives*—A clear statement of each activity and subactivity and the expected outcome of the activity and subactivity must be presented. These outcomes represent the objectives and each should be stated in behavioral terms that can be quantitatively measured.
- *Events*—The beginning and the ending point for each activity and subactivity must be identified. This identification should include the starting date and the completion date for the activity and the subactivity.
- *Effort*—Each activity and its component subactivities must have an estimated level of effort reported in terms of hours per week that are devoted to the activity by staff who are directly involved.

USING SQUARES AND OTHER SYMBOLS

Small squares or boxes can be used in the same way that bullets are used. It is somewhat easier to use bullets, since the typewriter period symbol or o can be used. Squares are a little harder, since one must be sure the lines are straight, and this cannot be done free hand. A ruler or Plotting Mark template can be used.

Interest and coherence can be added to a report by using both bullets and squares in different places. This method can be particularly effective if you use one symbol to designate one type of material and another symbol to designate a different kind of material. In effect, the two symbols become a form of coding that visually cues the reader mentally to categorize or differentiate the material. An example of this is as follows:

■ Prepare List of Objectives
  - Smith
  - Jones

- ■ Prepare Operational Definitions
    - Jenkins
    - Armstrong

What this example does is to cue the reader visually to the fact that boxes represent tasks, and bullets represent people.

Bullets and boxes can be used as solids (● ■) or open (○ □) in order to add further interest. Naturally, you can substitute any one of the variety of symbols shown earlier for bullets and squares.

SETTING OFF TABLES WITH BARS AND BOXES

Bars and boxes can be used to set off tables from the running narrative of a report so that the reader will give special attention to the table. The use of this technique is illustrated in the following Examples G, H, and I, in which the same table is presented in three different ways.

Example G: Narrative and Unlined Table

During the last decade the life expectancy for men and women at time of birth has improved in both the developed countries and the less-developed countries, as seen in the following table:

LIFE EXPECTANCY IN YEARS

	1970	*1980*
DEVELOPED COUNTRIES		
Men	67	68
Women	74	75
LESS-DEVELOPED COUNTRIES		
Men	51	55
Women	53	58

It should be noted that while the less-developed countries still have a shorter life expectancy, they have improved it at a faster rate than have the developed countries.

Example H: Narrative, Table, and Box

During the last decade the life expectancy for men and women at time of birth has improved in both the developed countries and the less-developed countries, as seen in the following table:

LIFE EXPECTANCY IN YEARS		
	1970	*1980*
DEVELOPED COUNTRIES		
Men	67	68
Women	74	75
LESS-DEVELOPED COUNTRIES		
Men	51	55
Women	53	58

It should be noted that while the less-developed countries still have a shorter life expectancy, they have improved it at a faster rate than have the developed countries.

Example I: Narrative and Table with Bars

During the last decade the life expectancy for men and women at time of birth has improved in both the developed countries and the less-developed countries as seen in the following table:

LIFE EXPECTANCY IN YEARS		
	1970	*1980*
DEVELOPED COUNTRIES		
Men	67	68
Women	74	75
LESS-DEVELOPED COUNTRIES		
Men	51	55
Women	53	58

It should be noted that while the less developed countries still have a shorter life expectancy, they have improved it at a faster rate than have the developed countries.

INSETS

An inset refers to the use of lines to set off a part of the text to which you want the reader to pay special attention. Insets are often employed to set off introductory paragraphs, ending summaries, and supplemental material. The material, because it is placed in a box, is set apart from the normal flow of the typewritten text; this serves to cue the reader to the fact that the boxed material is either different from or more important than the other text on the page.

The lines for the inset can be done either on the typewriter or by hand using a ruler. The placement of the inset box can be almost anywhere on the page: top, middle, bottom, sides. Using insets that are rather wide, as in Example J, makes it easier for typists.

Inset boxes may be done just with lines or may have some shading added to give them depth.

Example J: Using an Inset

The results of the study stress that our success as an organization seems to be based on four factors that we trust all workers will continue to exemplify in their work this year. Perhaps most important is the degree to which there is consistency among all levels of workers in the extent to which they understand and endorse our prime organizational mission of preservation of natural resources. The study showed that employees ranging from chief executive to maintenance workers listed this as the primary goal of the organization. Second was the fact that all levels of compensation were commensurate with the levels paid in other governmental and private agencies in the natural-resources management and recreation field. Third was the strong indication that workers felt positively about our promotion policies, which give priority to present workers over outside persons in being considered for job openings. Last was the impact of our team decision-making process that was initiated last year in which weekly meetings are held to discuss pending policy issues before final decisions are made by management.

Four Success Goals

- Agreed-upon goals
- Competitive salaries
- Internal promotion
- Shared decision-making

Example K: Using a Grid

DEPARTMENTAL AID PROGRAMS

DEPARTMENTAL PROGRAMS	TYPE OF BENEFIT				ELIGIBILITY		
	Contract	*Grant*	*Loan*	*Fed. Asst.*	*Public*	*Private*	*Other*
Community Action Programs	•	•			•	•	•
Economic Adjustment Program		•	•			•	
Commodity Distribution Program				•	•	•	•
Food Stamp Program		•			•		
Nutrition Research Support	•	•			•	•	
Code Enforcement Demonstration		•		•	•		
Local Development Loans			•		•	•	•
Education Assistance Program		•	•		•	•	•
Operating Loans			•			•	
Injury Control Program	•		•			•	
Environmental Sanitation		•		•	•	•	
Physical Fitness Program	•	•		•	•	•	•
Faculty Development Grants		•			•	•	
Library Assistance Program				•	•		

BORDERS

Cover pages or special pages within the report can be made interesting by adding a border design to the page. This is accomplished by using border tape, which comes in rolls or in packaged strips. They come in black and in a wide variety of designs and are self-adhering.

FIGURES AND DRAWINGS

Occasionally you may want to have some symbolic drawing in your report. There are many kinds of transfer sheets on the market with drawings of people, trees, buildings, and the like that can be obtained for this purpose.

LETTERING

Lettering that appears to be hand-done and relieves the monotony of the single-type style of the typewriter can be easily accomplished by use of transfer lettering sheets. These sheets contain letters and numbers in more than two hundred different styles and come in various sizes. They are applied either by burnishing the back to transfer the lettering to the typing paper, or the letters may be self-adhering and can be transferred by cutting them out and sticking them on the page. They are most useful for adding variety to headings and subheadings in the text.

GRIDS

A popular technique that is both interesting and efficient is the use of a two-way grid as a way of classifying and/or summarizing information. In Example K we show just one of the many formats that can be used. By varying the use of symbols and grid lines you can develop different types of grids for the presentation of categorized information in your report.

In this chapter we have covered just a few of the many different ways one can add interest and variety to the narrative report. With imagination and experimentation you can invent your own techniques by adapting, modifying, and extending the basic approaches that have been shown.

CHAPTER TWELVE

Resources

Chart-making has been made very much easier and accessible to the nonartist as the result of the availability of a wide spectrum of commercial products for the graphic arts. These products are produced by a variety of companies and are available in art, office-supply, and stationery stores or departments. Because of the large number of different products, most stores simply cannot stock them all, but they can order them for you from the catalogues of these companies. You can educate yourself in this field by browsing through stores and obtaining and looking through their catalogues and those of the producers and distributors of graphic products.

The main chart-making aids to be considered are:

- pens, pencils, and erasers
- templates
- graph paper
- transfer sheets for shading and hatching

- knifes
- lettering aids
- protractors, compasses, and triangles
- aids to correction
- charting tapes
- reference books

These and other resource materials are described in subsequent sections of this chapter.

PENS

Use only black "ball-point"-type pens, since these will reproduce the best. Get a variety of point thicknesses so you can vary the width of your lines. Tips come in nylon, metal, plastic, or fiber. Experiment until you find the ones that suit you best. Popular brands are Papermate's Flair Pen, Pentel, Berol, Pilot, and Sanford.

Do not use the wide-tipped marking pens or art markers, since these will give you lines that are too wide or lack sharpness of definition. Always be sure you get pens that have ink that dries instantly.

There are also technical drawing pens, which resemble fountain pens in that they have a reservoir to be filled with drawing ink. Do not use writing ink. Some pens have interchangeable points of many different widths. Brands include Koh-I-Noor, Faber-Castell, Staedtler, and Keuffel and Esser.

PENCILS

Pencils come in varying grades, from hard to soft. The harder the pencil the smaller the diameter of the lead used in wood pencils. For the kinds of work necessary for preparing the kinds of charts in this book it is more satisfactory to use a medium- or soft-grade pencil, usually designated as 2-B (soft) or 2-H (harder). The hardest grade to use is 4-H. Pencils harder than this one are too fine and may also tend to tear the paper. Use of mechanical pencils is not recommended. Faber, Eagle, Venus, and Dixon are popular brands.

When using the wood pencil always be sure that the point

is kept sharp, using a pencil sharpener, not a knife.

A handy pencil is the nonrepro or nonprint blue or purple lead pencil. The blue will not reproduce when making copies. These pencils are useful for putting in guidelines and for tracing or sketching before finishing with black ink. Staedtler, Faber, and Eagle make nonreproducible pencils.

☐ ERASERS

Use an eraser separate from the one that comes on the end of the pencil. Most pencil erasers tend to smudge. An art gum, pink pearl, or ruby eraser is best. Popular brands are Faber, which makes a wide selection, and Fiscol Company's ARTGUM. Erasing machines and erasing shields are unnecessary for this kind of chartmaking.

☐ TEMPLATES

There are hundreds of different templates available. Made of transparent plastic, a template has a number of cut-out shapes and symbols that serve as guides to prepare good charts with the help of templates. You can draw perfect circles, rectangles, and other shapes with a template. The average template costs about four dollars. If you are going to prepare reports frequently, it will be worth buying three or four different kinds in order to obtain the full range of shapes you may need. Unfortunately, among the hundreds of template styles on the market there is no single template that is suitable for use in preparing charts.

The most useful templates include the following:

Computer Diagrammer

Generally about 4″ or 5″ wide and 8″ or 9″ long, this template includes the shapes that are used in preparing computer diagrams, which are also very appropriate for use in preparing charts. Templates of this type are especially useful for making flow charts, organization charts, and bar charts. The range of shapes can be seen in Figure 9-1 in Chapter 9. Two of the edges of these templates are ruled in inches marked off in tenths, eighths, and sixths. The other edges include different types of brackets.

Perhaps the best variety of shapes is on the Pickett 1011 Computer Diagrammer. Koh-I-Noor, another major brand, makes a diagrammer template No. 830054 that has somewhat larger shapes than the Pickett, but not as wide a range of different shapes. Berol RapiDesign (R-54) and Charette (C-1892) make Flow Chart Symbol templates that are similar to the Pickett 1011 Computer Diagrammer and have good shapes for charts.

A similar template is called Standard Logic Symbols; it is not as well adapted as the Pickett 1011 Computer Diagrammer or Flow Chart Symbol types but will suffice if you can't obtain the others.

Arrows, points, and brackets are usually included in most computer-diagrammer and flow-process templates.

Giant Circles

Most of the difficulties in making pie charts can be solved with a template known as Giant Circles. Pickett, Timely, Charette, and Berol all have templates of this type. Be sure to get one that includes large circles of up to 3″ in diameter, such as Timely's 87 or Pickett's 1201. It is most helpful if you obtain one that has the diameters marked, since this will allow you to locate the center point easily and divide the circle into segments with no trouble.

Organization-chart Templates

A specialized template that includes a variety of rectangles and squares and some smaller circles and triangles is the organization-chart template, such as the Berol RapiDesign R-556. This template can be used for organization charts, flow charts, and bar charts. It also has arrows, brackets, and ruled edges that are useful.

Plotting-symbol Templates

The plotting-symbol template such as the Berol RapiDesign R-59 includes fifteen different symbols in graduated sizes. These symbols can be used in preparing line charts and can be adapted for use in adding interest and variety to the narrative layout of the report, as pointed out in Chapter 11.

☐ GRAPH PAPER

Problems of layout, alignment, and spacing are minimized or eliminated by drafting charts on lined graph paper. Lined in both horizontal and vertical directions, graph paper is comprised of equal-size squares and comes in a variety of squares to the inch. For most charts, the best size to get is four, five, or six squares to the inch. The paper comes either in pads or loose in packages. It is either opaque or of a semitransparent tracing quality. The tracing quality has two advantages. One is that you can use it to copy charts from other material. The other is that the lines will not reproduce. This means that you will not have to recopy a chart onto the plain white typing paper being used in the rest of a report in order to photocopy it or to reproduce it by other direct copying means.

Graph paper is available from most leading paper producers. Keuffel and Esser is a leading producer of graphic products whose papers are preferred by many professional draftspersons and illustrators. Alvin and Company makes a handy pad of nonreproducible paper called Alva-Line.

☐ SHADING AND TONING

In order to shade charts one can use shading films that come in an extremely wide variety of patterns, hatchings, and graduated tones from light to dark. These come in several different types, but the most satisfactory are those of the cut-out, self-adhering type. Select the particular pattern or tone you want to use and place it over the part of the chart you want to shade. Then, following a ruler edge, lightly cut along the edges, using a razor or a knife. Place on a piece of cardboard and cut the desired piece out of the sheet, remove the backing, and stick the piece on the chart.

Shading sheets also come in pressure-sensitive sheets. Cut out a piece slightly larger than the area to be shaded. Lift the piece off the sheet and place the piece in position on the chart. Rub the piece gently to fix it to the page, cut off any surplus, and then rub or burnish the piece to transfer the shading. Popular brands of these sheets are Zip-a-tone, Letratone, Normatone, and Formatt.

☐ KNIVES

An Exacto Knife No. 1 can be used to cut out transfer sheets. Mecanorma makes special knives, including their 61-6494 Corde Cutter, 56-1265 Swivel Knife, and 61-6491 Retractable Knife.

☐ SYMBOL AND PICTORIAL SHEETS

Similar to shading sheets, this type of transfer sheet includes drawings and other symbols. They are sometimes referred to as architects' symbols and include drawings of people, houses, landscape, trees, and vehicles, as well as a variety of designs. They can be used for preparing pictorial bar charts, report covers, and for adding interest to a report with an occasional drawing. Especially useful are sheets that contain all sorts of different arrows for use in flow charts, and those with various plotting marks for use in adding interest to report layouts.

☐ LETTERING

It is not necessary to do any free-hand lettering when lettering charts. Most of the lettering is done on the typewriter. In addition, there are several different types of lettering aids that can be used for headings, titles, and labels that are easy to use and will add variety and interest to the report.

One of these is transfer lettering sheets that are pressure-sensitive. These sheets of thin, transparent material contain letters and numbers of various styles and type faces. They are used by removing the backing sheet and positioning the letter over the drawing in the desired position. Then rub the back of the letter and it will adhere to the paper. Transferred letters can then be burnished to assure greater adhesion. Leading brands available from office-supply and art stores are Mecanorma, Letraset, Cello-Tak, and Chartpak. Once you have mastered lining up the letters by using guidelines, the pressure-sensitive lettering sheets are the best to use.

Another type is cut-out acetate letters that are cut out of a carrier sheet of acetate, lifted off, and placed on the artwork in the correct position. They are self-adhering.

There are also plastic lettering guides or lettering templates, but these are much more difficult to use satisfactorily than are the transfer sheets.

☐ BURNISHING

Art stores carry inexpensive wood, plastic, and metal burnishing tools that are very useful for rubbing transfer sheets and burnishing letters after they have been transferred in order to assure the permanency of the adhesion to the copy page.

☐ PROTRACTORS, COMPASSES, AND TRIANGLES

Protractors are plastic or metal half circles or full circles that have the degrees of the circle marked on them. (There are 360 degrees in a circle. Each degree is equivalent to .36% of the whole circle.) Protractors can be used to measure off the segments of a pie chart and will help assure accuracy in their size. There are also percentage protractors that have the percents already marked on the protractor and do not require converting degrees to percents.

A compass is a useful drafting tool for drawing circles if you do not have a template. One advantage of the compass is that it can be used to make circles of any size, whereas with templates you are limited to the circle sizes on the template.

A drafting triangle made from transparent plastic can be

helpful in drawing straight lines that are parallel to the edges of the page and in putting in guidelines to be followed for lettering or drawings.

☐ CORRECTION FLUIDS

Typewriter correction fluid is an excellent way to cover up penlines or errors that you want to eliminate on drawings. This white opaque fluid comes in small bottles with a brush to apply it over the typing or penline to be corrected. Popular brands are Delete, which is a water-based solution, and Sno-pake.

☐ CORRECTION TAPES

Another satisfactory way to eliminate unwanted lines is to use self-adhering correction tape, which comes in narrow widths in strips or in a dispenser roll. Pull out the length needed and place it over the line to be eliminated.

☐ CHARTING TAPES

These are rolls of tape that generally come in widths from 1/64″ to 2″ in a dispenser or package. They come in solid black (or colored) lines on a clear transparent background in matte or glossy finishes. They are self-adhering and are very easy to use for all straight lines on a chart. In addition to solid lines, they come in a variety of different vertical and diagonally striped patterns to use to differentiate different lines in a chart. They also come in decorative designs that can be used as border tapes to add attractive touches to the appearance of routine pages.

Brands include Cello-Tak, Letraset, Normatape, and others.

☐ MAPPING

Among the more advanced kinds of chart are statistical and other types of maps. These graphics are used to portray the organization of space, and spatial relationships and distributions. The simplest form of mapping is the spot or dot map, in which dots are superimposed over a map of an area to represent some distribution of persons. Preparation of most maps for report reproduction will often require some professional artwork in order to draw the map accurately.

☐ STATISTICAL AND OTHER REFERENCE BOOKS

Charts and graphs are widely used in the field of statistics both for purposes of displaying information and as methods of statistical analysis and problem solving. In addition there are many highly technical and complex uses of graphing and charting in the various scientific fields, mathematics, engineering, architecture, and the like. For the person interested in the statistical and more advanced aspects of graphic technique there are many, many introductory statistics books that any public library or university library will have available.

In addition, there are two books that deal specifically with statistical charts:

Peter H. Selby, *Using Graphs and Tables.* New York: John Wiley & Sons, Inc., 1979.
A survey of graphic methods primarily focused on the interpretation rather than the construction of graphs and tables. Organized as a self-teaching guide, it provides numerous examples of different charts, and questions and answers regarding their interpretation and statistical meaning. It covers line graphs of all types, area graphs, bar graphs, and special graphs of other types.

Calvin F. Schmid and Stanton E. Schmid, *Handbook of Graphic Presentation,* 2nd ed. New York: John Wiley & Sons, Inc., 1979.
A comprehensive survey of the innumerable variations in chart formats, primarily geared to the professional draftsperson, statistician, cartographer, and communications specialist. It has major chapters on drafting techniques, rectilinear coordinate charts, statistical maps, computer graphics, and projection techniques in graphic presentation.

For an interesting look at the way in which charts can be used to mislead a reader a popular book is:

> Darrel Huff, *How to Lie with Statistics* New York: W. W. Norton & Company, Inc., 1954.

☐ OTHER REFERENCES

A brief twenty-four-page pamphlet on charts has been published by the U. S. Government, which is titled *Descriptive Statistics: Tables, Graphs, and Charts,* and can be ordered from the U. S. Department of Health, Education, and Welfare, Public Health Service, Center for Disease Control, Atlanta, GA 30333.

In addition, a helpful book is:

> Mary Hill and Wendell Cochran, *Into Print: A Practical Guide to Writing, Illustrating, and Publishing.* Los Altos, Calif.: William Kaufmann, Inc., 1977.
> A manual that describes the basic requirements for preparing book manuscripts for publication, including how to prepare the graphics in proper form for publication.

One of the most elaborate examples of the use of graphics can be obtained from the U. S. Government. Titled *Social Indicators,* it was compiled by the Office of Management and Budget and is available from the U. S. Government Printing Office, Washington, DC 20401.

Interesting examples of persuasive graphs and charts appear in every issue of the *New York Times,* and you will find that many of the illustrations in the *Times* can be adapted for use in reports.

Don't overlook the annual reports of large companies, organizations, and agencies. They include extensive use of charts that are prepared by professional illustrators. You will find a number of examples you can adapt.

Finally, if you get in the habit of thinking graphically, you will find no end to the graphic possibilities that pop into your own head for use in your reports.

Index

ABOUT THE AUTHOR

Robert Lefferts, Ph.D., is Associate Dean for Academic Affairs and a professor at the State University of New York at Stony Brook, New York. He is the author of more than thirty-five major reports for governmental and private organizations. He has conducted training programs in proposal writing and report preparation at the State University of New York and for private organizations. He is the author of *Getting a Grant: How to Write Successful Proposals* (Prentice-Hall, 1978), a selection of the Macmillan Behavioral Science Book Club and the Mental Health Practitioners Book Club. For seven years he was Vice-President for a management consulting firm; he has been a consultant to various universities and federal, state, and local government agencies and has been an administrator of planning and research organizations.